American M
Physicians dedicated

Health & Safety Management
for Medical Practices

Evaluating Risk and
Implementing Safety for
Physician Offices

Linda Chaff

Health & Safety Management for Medical Practices

© 2002 by the American Medical Association

Printed in the United States of America.

All rights reserved.

Internet address: www.ama-assn.org

This book and CD-ROM are for information purposes only. They are not intended to constitute legal or financial advice. If legal, financial, or other professional advice is required, the services of a competent professional should be sought.

No part of this publication (book or CD-ROM) may be reproduced, stored in a retrieval system, or transmitted in any form or by any means electronic, mechanical, photocopying, recording, or otherwise, without the prior written permission of the publisher.

Additional copies of this book may be ordered by calling 800 621-8335. Mention product number OP209300.

ISBN 1-57947-083-1

BP38:0162-00:10/01

contents

Preface vii
Acknowledgments ix
About the Author xi
A Note from the Author xiii

1 Getting Started 1

Determining the Role of Management 2
Addressing Diverse Needs 3
Performing a Needs Assessment 4
Setting Priorities 9
Developing Policies and Procedures 9
Communicating a Commitment to Safety 10

2 The Creation of Occupational Safety and Health Agencies 13

Governmental Compliance Agencies 13
Other Federal Agencies 22
State Regulations 24
Nongovernmental Organizations and Agencies 25
Guidelines for Regulatory Inspections 29
Other Information and Assistance 31

3 Infection Control 33

Examples of Infectious Diseases 34
Infectious Disease Transmission 34
Defense Mechanisms 35
Infection Control Resources 35
Infection Control Program Elements 36
Needlesticks and Other Sharps Injuries 37
OSHA's Standard for Bloodborne Pathogens 41
Alternative Infection Control Methods 51
Occupational Exposure To Tuberculosis 52

4 Chemicals 59

Understanding Occupational Exposure Values 59
Chemicals Used In Medical Practices 61
Working With Chemicals 70
OSHA Hazard Communication Standard 73

5 Other Health and Safety Issues 79

Pharmaceuticals and Controlled Substances 79
Medical Gas Handling and Storage 82
Servicing Machines and Equipment (Lockout/Tagout) 94
Slips, Trips, and Falls 96

6 Medical Equipment Safety Management 99

Standards and Recommendations 99
Types of Medical Equipment/Devices Typically Found in Medical Practices 101
Procurement Planning 102
Elements of a Medical Equipment Safety Program 102
Lasers 108

7 Personal Protective Equipment 111

Program Components 111
Latex Glove Allergy 120
Respiratory Protection 122

8 Hazardous and Medical Waste Management 125

Categories of Dangerous Health Care Waste 125
Requirements for Environmental Management 126
The Process of Waste Management 126
Liability Issues 126
Hazardous Waste 127
Regulated Medical Waste 134

9 Promoting and Securing Employee Involvement 147

Establish Meaningful Safety and Health Committees 148
Integrate Personal and Family Health 149
Develop Promotion and Recognition Programs 153
Set Up Control Methods 154

10 Ergonomics 159

What Is Ergonomics? 159
Overview of the Problem 161
Overview of Ergonomics in the Workplace 161
Applying Ergonomics in a Medical Office 162
Cost Vs. Benefits 170
Resources Available for Ongoing Awareness 172
OSHA and Ergonomics 174

11 Workplace Violence 175

Risk Factors 175
Defining Workplace Violence 176
Types of Workplace Violence 177

Domestic Violence   183
Recognizing Risk Factors   184
Organizational Recovery After an Incident   188
A Workplace Violence Prevention Program   189
OSHA's Involvement   190

12 Indoor Air Quality   193

Overview of Indoor Air Quality   193
Types of Complaints   194
Potential Sources of Building-Related Illnesses   195
Identifying Building-Related Problems   196
Approaches for Solving Indoor Air Quality Problems   200
Maintaining a Healthy Indoor Environment   204

13 Patient Safety Program   207

Facility and Environment   208
Additional Considerations   213
Emergency Management   218
Medical Equipment Management   218
National Patient Safety Foundation   218

14 Emergency Management and Response   221

Emergency Management Concepts   221
Hazard-Specific Information   224
Emergency Planning Requirements   240

15 Education and Training   247

Setting Up a Safety Training Program   247
Planning for Successful Safety Training   248
Fundamentals of Safety Training   256
Regulatory Agency Training Requirements   257
Training Techniques   257
Training and Learning Evolutions   261
Levels of Training   263
Generation-Specific Training   267
Safety Training for Multicultural Employees   273
Auditing Training Retention   275

16 The Impact of Accidents and Illnesses   277

National Injury and Illness Reports   278
Reporting Injuries and Illnesses   279
Investigating Incidents   281
Eliminating Accidents Through Data-Tracking Programs   285
Workers' Compensation   285
Reporting Requirements   293

17 Safety Program Evaluation 301

　　Ongoing Program Evaulation 301
　　Annual Program Evaluation 303
　　Evaluation of New and Revised Safety Programs 304
　　Program Evaluation As a Tool for Resource Allocation 304
　　Benchmarking Safety and Health Programs 305

Appendix About the CD-ROM 307

Index 309

preface

The first occupational physician was the eighteenth century Italian, Bernardino Ramazzini. His world was often dark and filled with despair. Infectious disease etiology was poorly understood, and each day most of his fellow Italians struggled at harsh jobs to make a living for themselves and their families.

Beyond Ramazzini's incredible drive to ease the suffering of workers in all occupations, his most remarkable quality was being able to understand workers so well:

> *For we must admit that the workers in certain arts and crafts sometimes derive from them grave injuries, so that where they hoped for a subsistence that would prolong their lives and feed their families, they are too often repaid with the most dangerous diseases and finally, uttering curses on the profession to which they had devoted themselves, they desert their post among the living.*

Bernardino Ramazzini, Diseases of Workers[1]

Alice Hamilton, an American physician and pioneer in industrial toxicology in the nineteenth and early twentieth centuries, studied the effects of chemicals in the workplace. Hamilton was a social reformer who also understood the cost that work imposed on people who were only trying to provide for their families. Alice Hamilton served as an inspiration to those who worked to enact the Occupational Safety and Health Act of 1970.

Since the times of both Ramazzini and Hamilton, our lives have improved immeasurably. Occupational health and safety now manifests itself with worldwide determination and passion. Nevertheless, there is further improvement to be made because occupational injuries and illnesses still "...often come gradually and with stealthy foot."[1]

In past centuries, most jobs involved strenuous physical labor. Now, in the twenty-first century, most jobs in the industrialized world have shifted to service industries, including health care delivery. Society has traditionally viewed health care workers as free from the burden of occupational injury and illness, but we now know that even health care workers are the target of that "gradual and stealthy foot."

Because of the untiring work of people like Ramazzini and Hamilton, and the efforts of many public health officials to develop and enforce laws to protect human health and safeguard the natural environment, we now more easily recognize the effect that health and safety has on our daily lives and overall well-being.

We all sometimes "bring work home." We also "take home to work." All aspects of our lives are so intertwined that work and home are truly inseparable. It is no longer appropriate to leave our health and well-being in the hands of our employer or health care professional. Just as employers have a responsibility to provide a workplace free from recognized hazards, employees are responsible for working in a safe and healthful way at work and at home.

This book integrates individual health and well-being with occupational health and safety. Correspondingly, we introduce the concept of *Total Health and Safety*, which recognizes for the first time that individuals' chances for working safely on the job are vastly improved if they proactively maintain their personal health and well-being through diet, exercise, hygiene, and personal safety measures.

More importantly, this book presents a road map for ensuring that the journey from home to work and back again can be completed, so that every worker can do his or her job and support a family free from those debilitating injuries and illnesses described by Ramazzini.

Seventeenth century poet John Donne wrote:

No man is an island, entire of itself; every man is a piece of the continent a part of the main. . . .[2]

And so it is with our seemingly separate activities in life.

ENDNOTES

1. Ramazzini, Bernardino. *Diseases of Workers.* Translated from the Latin text *DeMorbis Articum* (1713) by Wilmer Cave Wright. OH&S Press 1993. NorthWest Training and Development: Thunder Bay, Canada.

2. Donne, John. *The Complete Poetry and Selected Prose of John Donne* (Modern Library Series). New York: Random House; 1994.

acknowledgments

My gratitude goes first to the American Medical Association who engaged me to write this book. In particular, J.D. Kinney. Also Marsha Mildred, who provided direction and persistent stimulation toward deadlines. Thank you J.D. and Marsha! Also, thank you Annette Testa for providing expert editing. And I want to express my sincere appreciation to Katharine Dvorak for all of her assistance throughout the course of this book.

With heartfelt appreciation, I also thank so many people whose untiring support and work made this book possible. Their experience, ideas, information, passion, and incredible knowledge contributed to the final product. I cherish their insights and support: Joe McFadden, CHSP, MT, who provided technical information, stacks of research, and contributed his years of experience in workplace safety; Linda Glasson, who continually showed me how to keep focused on the message; Patricia Meyer for her invaluable experience in workplace violence and critical incidents; Ed Savard, PhD, CIH, CSP, for his unique ability to design and market safety concepts and make programs enjoyable as well as practical and his eagerness to share those talents; Joan Hetzler, for her unique capability to help me communicate more clearly and maintain my writing style (Joan is my most valued editor); Yvonne MacManus, who used her experience in teaching and writing to help layout chapter concepts and explore unique presentations for the material; Dee McCamish and Denise Reed, my colleagues who bring special contributions and never run out of questions, picky editing, or suggestions.

I would also like to thank Terry Day, my cousin who provided information about special services in medical practices; Tommy Gladson's experience with state drug diversion teams and health care facilities; Al Kuntz for his extensive background and knowledge in medical equipment safety; Ray Mulry, PhD, my good friend who has written many books on the key elements of health and well-being; Barbara Ondrisek, who taught me how to wash my hands. I invariably think of Barb and her clever demonstrations on infection control during defined moments; Ron Wene at Guardian Protection Systems whose expertise in emergency management is always such a help to me; Fred Burt, PA; Marlene Hardaway, RN; George Mara, Gene Redden, Lee Shratter, MD, Priscilla Thal, and the folks at my neighborhood CVS Pharmacy.

I also developed new relationships with many public health officials and other people whose companies, agencies, and medical practices provided figures, tables, and critical information throughout the book. I appreciate their cooperation and interest in this product. If I do list names, I run the unforgivable risk of leaving some people out.

And of course I am forever beholden to my husband, Bill, and my son, Brian, whose support and patience are never ending and who have *unconditionally* learned to accept the time it takes to produce books.

the author

Linda Chaff is president of Chaff & Co, a corporate communications firm in Chattanooga, Tennessee, where she develops ideas, training concepts, promotional campaigns, and programs to improve individuals' skills to make an exciting difference in the results and growth of companies and organizations. Her credits include the Russell L. Colling Literary Award, and appointments involve committee and board positions with safety and civic councils including the executive committee of the National Safety Council's Health Care Section.

Chaff & Co's creative teams of writers, designers, and industry experts write and produce materials on health and safety, customer service, and training techniques. Linda is also the author and producer of numerous books, interactive presentations, training manuals, and plays and productions published and presented by the American Hospital Association, the National Safety Council, state health care associations, medical facilities, and private industry.

When developing training programs using her books, Linda recognizes that a boring presentation assures a swiftly forgotten lesson. When preparing for training sessions, Linda combines drama with humor, plus active and inactive participation. Just as timing is everything in theatre, so it is with workplace health and safety programs.

a note from the author

I grew up on a farm in Concrete, Washington. Farming was a difficult way to support a family, so my father, Joe Styles, worked in a sawmill during the day to help supplement family income. Before there was an Occupational Safety and Health Act (OSHA), he experienced the trauma of a debilitating injury at the sawmill, losing two fingers on his left hand while operating an unguarded saw to cut heavy logs. This experience, and the stories about my grandparents who suffered disabling workplace injuries, set me on the same journey as Bernardino Ramazzini—to try to ensure that everybody's husband, father, grandfather, wife, mother, and child can be in a workplace that is healthy and safe.

Now I have come to realize that traditional workplace safety and health management is only a part of the solution for ensuring worker protection. Whether we are discussing the health care industry or the more traditional workplace, I have learned that working safely on the job is vastly improved if people maintain their personal health and well-being. This idea of total health and safety is the core of this book.

I hope you enjoy and benefit from reading *Health and Safety Management for Medical Practices*. If you have any thoughts or experiences you would like to share, please contact me; I would be delighted to hear from you.

Sincerely,
Linda Chaff

Linda Chaff
Chaff & Co
600 Republic Centre
Chattanooga, TN 37450
423 266-5541
www.chaffco.com

chapter 1

Getting Started

People are the heart of safety programs.
Linda F. Chaff[1]

Today's medical practices have ever-escalating responsibilities. They must ensure the survival of the practice through sound financial management, meet the competitive challenges of the industry through patient services, and provide a safe and secure environment for employees, patients, and others.

Until recently, medical practices focused almost exclusively on maintaining the health and safety of patients, not employees. Historically, medical practices have been considered safer than other work environments, and occupational injuries and illnesses among medical practice staff were often unrecognized. Factors that contributed to this include:

- The traditional view that health care employees were capable of maintaining their health without assistance.
- The availability of informal consultations with physicians appeared to reduce the need for formal safety and health programs.
- Medical practices were traditionally geared toward treating patients.

Infectious diseases were among the first recognized health risks for patients. As medical practitioners struggled to combat those diseases by improving patient health care delivery, there was a simultaneous benefit of improved employee health. Some of those improvements involved basic sanitation. For example, Florence Nightingale found that opening windows and introducing fresh air improved patient recovery. Similarly, Ignaz Philipp Semmelweis, the Austrian surgeon, observed that routine hand washing resulted in improved surgical recovery.

At the beginning of the twentieth century when physicians began working with X rays and potentially explosive anesthetic gases, such as ether, health-care practitioners began to recognize the potential for safety and health hazards among their employees from sources other than infectious diseases. The recognition of these new hazards called attention to the many dangers that face health-care workers. From that point on, medical practitioners began to monitor their employees not only for infectious diseases, but also for other safety and health problems associated with their work.

Hospitals began to establish formal safety programs, an effort that is now continuing in individual medical practices. The following are the various steps needed to establish a successful safety program in a medical practice:

- Determining the role of management
- Addressing diverse needs
- Performing a needs assessment
- Setting priorities
- Developing policies and procedures
- Communicating a commitment to safety

DETERMINING THE ROLE OF MANAGEMENT

The role of management when establishing safety programs in medical practices includes: establishing a vision, integrating safety, and committing to the program.

Establishing a Vision

In general industry, there is a direct correlation between operational excellence and safety excellence. Conversely, it has been established that organizations that are unable to manage their safety programs are also unsuccessful at managing other aspects of their business. To increase competitiveness, bottom-line performance, and employee morale, safety must be addressed.

A strong comprehensive, profitable safety program is manageable, continuing, and equal to other business objectives in reducing costs, increasing service, improving product quality, and enhancing employee relations. This means that if the vision of employee safety and health is established and integrated into patient health care delivery, medical practices will recognize the same benefits.

Safety is a journey, not a destination, and must be established as part of the practice's values and culture.

Integrating Safety into the Medical Practice

When safety becomes a routine part of the medical practice vision, employee illnesses and injuries decrease and productivity and effectiveness increase. If safety awareness keeps pace with advancements in patient health care delivery, then the quality of patient care improves and the practice reduces its cost. To establish a successful program, the following three core safety program elements are essential:

- *Human values.* Employees should be able to come to the medical practice, work in a safe environment, and return home safely.
- *Business.* Safety is an integral part of running a business and should be equal to other business objectives, such as improving quality, reducing costs, and improving patient relations.
- *Regulatory.* The Occupational Safety and Health Administration (OSHA), the US Environmental Protection Agency (EPA), and other agencies prescribe safety requirements and program elements for medical practices.

Human values are clearly the most important basis for the safety program. Because medical practitioners care about people, the development and promotion of a vision of safety excellence should be forthright. It benefits employees, employees' families, patients, and the medical practice, and creates employee confidence in management and the practice. With confidence in place, the medical practice can then achieve other objectives, such as reduced costs, increased revenue, higher quality, enhanced employee relations, and improved morale. The employee's motivation to improve his or her performance in safety is a key element of the safety program.

Committing to the Safety Program

Many medical practices want to be recognized as leaders in the delivery of patient care, and they can also be leaders in employee safety and health. Part of making this commitment to safety and health is understanding the value of safety excellence and knowing that it is something that employees achieve, not something provided by the employer. Instead of focusing only on providing a safe place, medical practices must also focus on providing an environment in which employees eagerly participate in maintaining a safe work environment.

The willingness of employees to actively participate in a safety program is the second key to a successful safety program. Actions of people, not workplace conditions, cause the vast majority of workplace injuries. Therefore, it is essential that employees are motivated to improve their safety performance.

Safety Program Objectives
Safety program objectives must include improvements in safety awareness, safety practices among employees, and safety policies that relate to the employees' families. The addition of a safety program to the medical practice does more than just reduce workers' compensation costs. It catalyzes team efforts in all aspects of the practice's operations, boosts morale, and increases cooperation among employees.

Safety Program Outcomes
Outcomes of commitment to a safety and health program include enhanced awareness on the part of employees about good work practices and a renewed focus on changing their individual work and health habits.

ADDRESSING DIVERSE NEEDS

The diverse safety and health concerns in medical practices are traditionally divided into hazards that pose an immediate threat and those that cause long-term problems. Safety hazards include sharp-edged equipment, electrical current, and dangerous floor surfaces, which can contribute to slipping, tripping, or falling. Health hazards are often more difficult to identify. They may result in an immediate illness or in the long-term development of disease. Although a needle puncture may result in hepatitis in 90 to 180 days, exposure to excess radiation or to some chemicals may not pose any noticeable health effects for years or decades. Therefore, employees may appear and feel healthy when, in fact, their health is being slowly or silently compromised. Because employees are often exposed to hazards from chemicals that are newly introduced into the medical practice, the effects of their exposure may not be well known. For this reason, employees may have difficulty associating a new illness with workplace exposures.

PERFORMING A NEEDS ASSESSMENT

An assessment provides the medical practice with specific information that is needed to decrease worker injury and illness, reduce or eliminate workers' compensation claims, build employee morale, and improve the overall operation of the practice. Once the needs assessment has been completed, areas for improvement can be identified and a comprehensive program that incorporates a time frame and priorities can be developed. The assessment process is the first step toward a comprehensive safety program that favorably affects the practice's financial stability, efficiency, and health care delivery.

Preparing for the Needs Assessment

A medical practice can conduct a needs assessment by using outside resources, such as consultants, or in-house resources. The advantage of using a consultant is that people who are familiar with occupational safety and health programs, as well as federal, state, and local regulations, perform the needs assessment. They also have first-hand knowledge about successful safety and health programs that are in place at other medical practices. Additionally, assessments performed by consultants tend to be more objective than those done in-house.

Some disadvantages to using outside resources include the lack of ownership of the assessment and the fact that the facility may not gain as broad an understanding of problems as it would if the survey were performed internally. These disadvantages can be overcome by having an internal team work with the outside resource.

If internal staff performs the assessment, the survey team should include the safety coordinator, union representation (if applicable), management, and other employees. Include individuals with specialized expertise (eg, ergonomics, infection control, fire prevention) as needed. The internal assessment process helps develop cooperative relationships and the teamwork that is essential to making changes to the existing safety program.

The survey team should be trained in the goals and objectives of the assessment and should agree on the approach to be used. It should survey employees for their perceptions of safety support and any elements of the safety program that might be deficient, such as training or incident follow-up.

A checklist is a useful tool when conducting the needs assessment, and OSHA, the National Institute for Occupational Safety and Health (NIOSH), and the Joint Commission on Accreditation of Healthcare Organizations (JCAHO) documents can provide invaluable assistance in creating that list.

A comprehensive needs assessment includes a review of the facility's pertinent records. These records give the survey team a sense of the medical practice's current safety program direction, including past accomplishments and oversights. Records that should be reviewed include:

- Incident reports
- Safety committee minutes
- Government agency inspection results, including those from OSHA and fire departments
- Insurance company inspection records
- Workers' compensation statistics
- Policies and procedures
- Employee training records
- Internal audit records

As part of the preparation for the assessment, these and other records from at least the past 12 months should be gathered and summarized by the safety coordinator, saving the team time by not having to review past reports. The documents should include statistics and trends, such as the percentage of back injuries, needle sticks, slips and falls, and workers' compensation claims. If these documents are not available, it is apparent that the medical practice's information management system is not in place.

Conducting the Assessment

Once the medical practice has made the necessary preparations for the assessment, it is time to commence with the process. In order for an assessment to be both thorough and successful, it should include the following elements:

- Physical condition of the facility
- Employee perception
- Safety program in place
- Condition of equipment
- Employee training
- Hazard evaluation
- Medical surveillance

Physical Condition of the Facility

Assessing the physical condition of the facility includes hazard identification using resources, such as OSHA regulations, National Fire Protection Association (NFPA) life safety and health care facility codes, and the National Electrical Code (NEC). Hazard identification involves not only recognizing the hazards themselves, but also learning their specific characteristics and identifying the population at risk so that control programs can be designed. Following are examples of items that should be included on the physical facility checklist:

- Do storage practices or places create a fire hazard?
- Are exit lights functioning?
- Are ceiling tiles in place?

- Are fire extinguishers inspected and in place?
- Are stairwells used as storage areas?
- Are exits blocked?

Survey team members should communicate with employees who are familiar with the particular areas of the practice, follow a checklist, and ask any additional questions that may arise (eg, common health problems, hazards not on the checklist, any differences from a typical department in another practice).

Employee Perception

Employees at all levels are often uniquely aware of the problems that exist with their safety program. In many cases, these employees can suggest solutions to specific problems. Their suggestions are valuable to the team as it prepares recommendations for corrective action. In addition to soliciting employee perceptions about specific problems, the assessment team should survey employees to learn their perceptions of the overall safety climate within the workplace.

The survey can be done by either interviewing employees or asking them to complete a written form. With both approaches, employees should be guaranteed anonymity. The following is a list of suggested questions that can be included in the perception survey:

- Is the safety program perceived as positive?
- Are supervisors involved in accident prevention?
- Is safety considered important, and is good safety performance recognized at all levels of the organization?
- Do managers and employees communicate freely on safety issues and meet to formulate behavior-oriented safety goals?
- Do employees receive comprehensive safety training?
- Are safety inspections regular and effective?
- Does the practice take a strong, but fair, approach to safety rule infractions?
- Is there an effective system for dealing with reported hazards?
- Is the workplace climate conducive to adopting safe habits and work attitudes?
- Is there anything in the job that might affect the employee's health or the safety and health of other workers?

Many other types of questions may be included in a perception survey, and each practice must carefully tailor the survey to its own needs. Once complete, the survey results provide management with the information necessary to make changes in policies and procedures and with the justification that management needs to make changes in the safety environment.

Safety Program in Place

In addition to surveying employees for their perception of safety in the workplace, the assessment team needs to evaluate the effectiveness of the safety program that is already in place. To do this, the team needs to review records to determine the following:

- Does a safety committee routinely meet?
- If a safety committee does routinely meet, are union members (as appropriate), employees, and supervisors on the committee?
- Are there data-tracking records showing that safety problems are corrected in a timely manner?
- Are there current, easy-to-understand written policies for safety requirements?
- Is required training up-to-date?
- Has the effectiveness of the training program been reviewed and action taken to correct deficiencies?
- Are training materials recent and in compliance with regulatory requirements?
- Is there an effective program that treats and manages employee injuries?
- Is there a workers' compensation program?

Condition of Equipment

The survey team also should determine whether a system of preventive maintenance is in place. This can be done by reviewing selected maintenance records. The goal of the review is to determine the following:

- Is there a policy and procedure that addresses equipment use, failure, repair, and reporting?
- Is there an "alert" or recall program in operation for defective equipment?
- Is defective equipment repaired or discarded?
- Is broken equipment being used?

Employee Training

Employee training is an important element in the needs assessment. Policies and procedures are of little value if employees do not receive or retain the information they need in order to work safely. The assessment team should choose a representative cross section of employees and test their knowledge of key elements of facility and department safety programs. Testing should include talking with employees to determine their understanding of specific regulations, such as the OSHA "Right-to-Know Law" and the Bloodborne Pathogens Standard. Examples of the knowledge that employees should possess include:

- Where material safety data sheets (MSDSs) are located.
- What hazardous chemicals are used.
- What tasks involve exposure to bloodborne pathogens.
- What action should be taken in the event of exposure to blood or body fluids.

Testing can be reduced if records clearly show that employees have been recently trained and tested. If records cannot be found or are poorly maintained, the team must spend more time assessing the extent of employee knowledge.

Hazard Evaluation

Once hazards have been identified, they should be evaluated to determine their seriousness and to what extent they need to be controlled. Some hazards should be evaluated initially by an industrial hygienist, a professional whose assistance can be obtained from NIOSH, OSHA, private consultants, or, in some cases, insurance companies.

An industrial hygienist may take a variety of samples to help determine the extent of a workplace hazard. Most methods for chemical sampling require a laboratory analysis, which an accredited laboratory should perform. Such samplings measure contamination of work surfaces, contaminants in the employee's breathing zone, and exposure to chemicals and other environmental stressors.

After implementing controls, periodically check them to ensure that they are being maintained and that they are adequately protecting the employees. Prepare a chart or grid to list hazardous materials found in the medical practice, as well as where they are usually found, exposure limits, precautions to follow, and other relevant factors. This chart can serve as a quick reference and a means of tracking program development.

The signs and symptoms that employees report should be medically evaluated, taking care to avoid preconceptions about which ones are or are not work related. Occupational histories help evaluate the long-term effects of exposures. Biological monitoring may be an effective means for assessing past or current exposure to some chemicals.

Medical Surveillance

Appropriate medical procedures exist to evaluate the extent and the effects of some workplace exposures. For example, workers exposed to gluteraldehyde, the active component of many cold sterilants, may experience dermal or respiratory irritation and possibly sensitization. Other chemicals or their metabolites can be detected in an exposed worker's urine or blood.

The medical surveillance program should be designed based on information obtained from MSDSs, the industrial hygiene evaluation, and the needs assessment. The program should include required elements, such as consent and confidentiality, record keeping, and preplacement physical examinations. Also, consider exit physical examinations, as they may provide additional valuable information concerning the effectiveness of the health and safety program.

Applying the Assessment Results

The assessment results enable the medical practice to focus on immediate, as well as short- and long-term projects designed to eliminate existing safety problems and to anticipate future concerns. Written policies and procedures can help direct efforts and achieve safety goals. In addition, the assessment can foster a new relationship among administrators, managers, and employees. Because of ongoing interaction among the assessment team and the practice's staff, each individual understands that they are included in the decision-making process, building employee commitment to the safety program.

SETTING PRIORITIES

Once potential exposures and safety problems have been identified and evaluated, priorities should be established for controlling the hazards. All identified safety hazards should be promptly addressed and educational programs on subjects such as correct lifting procedures, infection control, and handling electrical equipment should be developed. Employees with potential for exposure should be informed and trained to avoid hazards, and controls should be instituted to ultimately prevent the exposures.

Once all information is gathered from the needs assessment, a report should be generated and submitted to management. A follow-up meeting should then be held to review the findings and to set broad priorities. Do not disband the survey team until priorities are set and approved.

Setting priorities for safety performance improvement is similar to any other process in business. The severity, frequency, and effect on employees are key factors to be considered. The prioritized results serve as the basis for development of short- and long-term goals, which often prove useful in measuring the safety program's success. Additionally, seeing achievable short-term goals on paper contributes to the participants' feeling of accomplishment.

DEVELOPING POLICIES AND PROCEDURES

After identifying priorities and setting goals, the next step is developing or revising the policies and procedures that address the goals. Policies form the heart of the safety program. They outline the facility's commitment to the program, define responsibilities, and provide details for the development of procedures, a primary resource for safety training.

Policies and procedures are evidence of the medical practice's intention to provide a safe environment for employees and patients. Additionally, these documents can be useful in defending the medical practice in the event of litigation. Moreover, JCAHO requires policies and procedures for accreditation, as do government agencies, such as OSHA.

One of the responsibilities of the safety coordinator and the safety committee can be to write and review policies and procedures. Policies must be developed for each major safety element (eg, fire prevention, incident reporting).

Policies and procedures should be approved in writing by the senior management of the medical practice and should be reviewed at least annually for effectiveness, employee reactions, regulatory developments, and changes in operations. Policies and procedures should be consolidated into a manual, which also includes the program goals, the names of safety committee members, and the statement of management's role and support. This manual can then serve as the comprehensive resource for the safety program.

Recordkeeping

One purpose of recordkeeping is to provide information that is important to ascertaining the effectiveness of the health and safety program. Keeping clear, concise yet comprehensive records on all aspects of the medical practice is essential. In a dispute, the practice's records will be opened for inspection. If they are not accurate or current, it could prove costly.

In the area of workplace health and safety, records also provide significant evidence that training has occurred. Recordkeeping is more than filing away receipts of manufacturers' warranties. Records are a means of keeping track of what has been done in the medical practice including upgrading equipment or procedures, establishing new policies based on regulatory changes, or highlighting the type of training that has been provided.

Employee attendance records are a valid means of keeping track of sick days, documenting holidays, or recording vacations. They can also be used to indicate problems in health and safety. For example, attendance records can be summarized to identify patterns or trends with employee sick days. If several employees have been out during a certain period of time for stomach ailments compared to the previous six months, could there possibly be a contributing health hazard in the workplace? Are employees reporting off from work due to sharp, shooting pains at their thumbs and wrists (symptoms of early carpal tunnel syndrome)? If so, attention should be given to the height and placement of computer equipment, including monitors.

Although recordkeeping is an invaluable means of maintaining or improving health and safety at the workplace, there can be many pitfalls, like permitting the process to dominate an employee's time at the expense of other duties and allowing recordkeeping to fall behind. If records are not regularly updated, they are not accurate. Furthermore, when the employee attempts to finally update the system, the task can seem too formidable, causing the employee to abandon the efforts.

Safety committees can be helpful in maintaining safety records, keeping proper documentation, and ensuring that corrective actions on major safety problems are taken in a timely manner. Safety records should include:

- Incident reports
- Internal audit reports
- Safety committee minutes
- Copies of external reports, such as fire department inspections and other government agency reports
- Workers' compensation records
- Records of employee training

COMMUNICATING A COMMITMENT TO SAFETY

A desirable safety environment is one in which people at all levels of the organization know that they are valued and are willing to put in the significant amount of effort that is needed to change the environment in which they work. Employees need to understand the logic behind the changes in safety performance and must be involved in the process of bringing about those changes. They need to be told hard facts about how poor safety performance effects the medical practice's ability to provide sound patient care and how failure to make the necessary changes will personally affect them. Employees want to know what new requirements are expected of them and, most importantly, how they will be involved.

Once change is initiated, it must be ongoing. Because employees constantly struggle to sort out which ideas and instructions to take seriously, top management's message about the medical practice's intentions regarding safety must be clear and unequivocal.

Other ways to communicate a commitment to safety is through training, resources, and budgeting. Personnel trained in occupational safety and health are needed to design, implement, and manage a safety program. There are many courses specifically designed to train nurses, safety officers, physicians, and other employees to recognize, evaluate, and control new and existing hazards.

Once the foundation of a good safety program is laid, a strong message of commitment to health and safety in the workplace will be assured.

ENDNOTE

1. Chaff LF. *Safety Guide for Health Care Institutions*. 5th ed. Chicago: American Hospital Publishing, Inc; 1994.

chapter 2

The Creation of Occupational Safety and Health Agencies

It is but a poor profit that is achieved at the destruction of a man's health.
Bernardino Ramazzini, MD[1]

Bernardino Ramazzini (1633–1717), an eighteenth century Italian physician, is often called the "father of occupational medicine." In his book titled *Diseases of Workers*, Ramazzini described the health problems he observed among the working men and women of Italy. Among the trades he wrote about were midwives, corpse bearers, and healers. Today people in those occupations are classified as health care workers. Ramazzini recognized that people are often unaware of the effect an occupation has on their health until it is too late. He wrote that the diseases he observed often came, ". . . gradually and with stealthy foot."[1]

Occupational health came late to the United States. Alice Hamilton (1869–1970) was an American physician and pioneer in industrial toxicology and is considered the "mother of occupational health and safety" in the United States. Hamilton was actively involved in social reform, supporting diverse causes including the enactment of worker compensation laws. Born shortly after the Civil War in New York City, she lived long enough to see the Occupational Safety and Health Administration (OSHA) become a reality.

Today, there are regulatory agencies, volunteer agencies, support organizations, and research agencies to deal with occupational disease and safety. Governmental compliance agencies can legally inspect, cite, and fine workplace infractions. Nongovernmental organizations and agencies lend their expertise to governmental compliance agencies. Most occupational health and safety agencies conduct research and provide guidance to improve health and safety at work through public health measures, industrial legislation, and so forth.

While some diseases still "come gradually and with stealthy foot," the work of these agencies and organizations has greatly improved unhealthy workplace practices involving safety, illness, and disease.

GOVERNMENTAL COMPLIANCE AGENCIES

There are many government compliance agencies that were put into place to address occupational health and safety in the workplace. The main agencies include: Occupational Safety and Health Administration (OSHA), the US Environmental Protection Agency (EPA), the US Food and Drug Administration (FDA), US Department of Transportation (DOT), Drug Enforcement Administration (DEA), Office of Justice Programs (OJP), US Nuclear Regulatory Commission (NRC), and Americans with Disabilities Act (ADA).

Occupational Safety and Health Administration

As the twentieth century progressed, it became clear that occupational injuries and illnesses could be prevented. While some worker protection laws were enacted at the state and local level to assist with specific industries or diseases, there was no single federal law directed toward improving the general health and welfare of all workers.

In response to a long history of serious workplace injuries, fatalities, and debilitating illnesses, and in recognition of the economic impact such problems had on American commerce, the federal government moved to establish uniform rules and regulations to govern American workplaces to ensure worker health and safety. In 1970, OSHA was established within the US Department of Labor and has become the centerpiece of an extensive array of public health agencies and organizations.

To achieve their mission of enabling all workers to return home safely everyday, OSHA established performance goals to guide the development of programs and activities. These include improving workplace safety and health, changing workplace culture, and securing public confidence in OSHA's programs and services.

Customer interactions and staff professionalism are also among OSHA's top priorities. Customer outreach programs include:

- An extensive publications program. Many can be downloaded from OSHA's website.
- Enhanced training, education, and assistance outreach efforts including technology-enabled training. These advances in communications technology promise to help OSHA enhance its education and compliance assistance activities.
- Computer/web-based training and satellite teletraining initiatives that allow OSHA to deliver more training to employers and workers at their job sites.
- Web-based Expert Advisors and Technical Advisors to help users identify workplace hazards that are unique to their situation.
- Electronic complaint filing for workers.
- A Voluntary Protection Program (VPP) designed to recognize and promote effective safety and health management.
- A Strategic Partnership Program (SPP) that is a voluntary, cooperative relationship with groups of employers, employees, and employee representatives to encourage, assist, and recognize efforts to eliminate serious hazards and to achieve a high level of worker safety and health.
- Special liaison to assist victims and families of workers who are hurt or killed on the job.
- Standards that are written in plain-language format to make them clearer and easier to understand.

- Public participation, submitted via the Internet, in the standards development process.

Although some businesses are not covered by OSHA regulations, the presence of the agency has raised people's expectations about the health and safety of their workplace. More than 100 million American workers and their families and the more than 6 million businesses that employee them are either directly or indirectly affected by OSHA activity.

By their very nature, medical practices are well aware of the dangers of infectious disease. However, some medical practices may not be aware of many of the safety hazards that exist within the clinical practice. For example, a pediatrician's office is likely to have safety considerations that are much different than those of a cardiology practice. Likewise, employees who are working in an orthopedic clinic may have a greater appreciation of general safety than staff in another type of medical practice.

The Occupational Safety and Health Act contains a General Duty Clause that summarizes an employer's responsibilities. Briefly, this clause says that employers are required by law to provide "employment and a place of employment which are free from recognized hazards that are causing or are likely to cause death or serious physical harm [to its employees]."

While some OSHA regulations specifically address medical offices, medical laboratories, and other related facilities, regulations can have a range of general applications. For example, OSHA enacted the Hazard Communication Standard in 1983 to help control exposure to hazardous chemicals and to ensure that workers had access to important information concerning the effects of exposure. Later dubbed the "Right-to-Know Law," it called for all employees to be informed about the hazards of the chemicals with which they work and how best to protect themselves. Because many OSHA standards and guidelines pertain to medical offices and laboratories, they could lead a practice into a smoother, more profitable operation.

Because OSHA is a federal agency with the authority to write citations, issue penalties, and impose civil or criminal fines for noncompliance, it enforces its regulations with on-site inspections. OSHA inspections, which include medical offices, laboratories, and related facilities, may be scheduled or unscheduled and are usually conducted because of:

- Probable, immediate danger
- Complaints from staff or others
- Fatal accidents or potential catastrophes
- Planned inspections for industries that are known to be extremely hazardous
- Follow-up inspections to ensure violations are corrected

In some states, it is not the federal OSHA that regulates workplace safety and health but a state-run occupational safety and health administration that operates under approval of federal OSHA. These states are referred to as State Plan states. If a state's economy is based in a particular type of industry, such as fishing, dairy farming, or coal mining, more stringent regulations may be required to meet the needs of that specific industry. Although these state agencies must meet the standards of the federal agency, they may tailor regulations to the most prevalent type of industry within their borders.

OSHA has long recognized that it cannot achieve its goals through enforcement alone. Thus, OSHA must reach out and partner with employers and employees to find, teach, train, and inspire people to safer ways of working.

US Department of Labor (OSHA), 200 Constitution Avenue, Room N3647, Washington, DC 20210. Telephone 202 693-1999. Website www.osha.gov.

US Environmental Protection Agency

The mission of the EPA is to protect human health and to safeguard the natural environment—air, water, and land—upon which life depends. The EPA writes regulations and guidelines to control the release of harmful matter into the environment.

What was once thought environmentally impossible is now not only possible, but also has already occurred. As trees died at higher altitudes from polluted air, respiratory ailments continued to increase, waters became contaminated, and fish became inedible, people realized that something had to be done to protect the environment. Many environmental ideas first crystallized in 1962, and the EPA was established in 1970.

The EPA's accomplishments affect medical offices in several ways, including the monitoring of air quality standards, controls on disposal of toxic chemicals, tight regulations involving pollutants used in a medical practice, banning of chemical compounds with the potential for causing cancer and other adverse health effects, and controlling the quality of public water supply. Anything that can have a negative effect on the environment is the EPA's domain.

The EPA's rules and recommendations divide substances into categories of possible or probable harm to health and the environment. They provide regulations for handling these substances, oversee the procedures of waste disposal sites, and instruct how to handle environmental accidents, such as leaks or spills.

OSHA regulates its Hazard Communication Standard, which deals with employee use and handling of hazardous chemicals while on duty. The EPA is concerned with how these chemicals affect the environment, both their handling and disposal. The enactment of major new environmental laws and important amendments to older laws in the 1970s and 1980s greatly expanded the EPA's responsibilities. The EPA now administers 10 comprehensive environmental protection laws, enacted by Congress, through which the EPA carries out its efforts. They include:

- Clean Air Act (CAA). Passed by Congress in 1970; additional amendments in 1990.
- Clean Water Act (CWA); passed in 1948.
- Safe Drinking Water Act (SDWA); enacted in 1974.
- Comprehensive Environmental Response, Compensation, and Liability Act (CERCLA or Superfund); enacted in 1980.
- Resource Conservation and Recovery Act (RCRA); passed in 1976.
- Federal Insecticide, Fungicide, and Rodenticide Act (FIFRA); enacted in 1947.

- Toxic Substances Control Act (TSCA); enacted in 1976.
- Marine Protection, Research, and Sanctuaries Act (MPRSA); enacted in 1972.
- Uranium Mill Tailings Radiation Control Act (UMTRCA); enacted in 1978.
- Pollution Prevention Act (PPA); enacted in 1970.

For more than 25 years, the EPA has worked to protect public health and the environment, ensuring that all Americans are protected from significant risks to human health and the environment where they live, learn, and work. Information or guidance may be obtained by contacting the EPA.

US Environmental Protection Agency, 1200 Pennsylvania Avenue, NW, Washington, DC 20460. Telephone 202 260-4048. Website www.epa.gov.

US Food and Drug Administration

Few governmental departments are more closely allied to a physician's medical practice than the Food and Drug Administration. The FDA Modernization Act of 1977 affirmed FDA's public health protection role and defined the agency's mission. Included in the mission is the responsibility for overseeing the safety of food and drug products, medical devices, and electronic products that emit radiation. The FDA also ensures that these products are honestly, accurately, and informatively represented to the public. The FDA promotes the public health by promptly and efficiently reviewing clinical research and taking appropriate action on the marketing of regulated products in a timely manner. With respect to such products, the FDA protects the public health by ensuring that drugs are safe and effective, there is reasonable assurance of the safety and effectiveness of devices intended for human use, and public health and safety are protected from electronic product radiation.

Under the FDA's requirements, medical practices must assume corrective action to assure the safety and well being of patients if a hazardous substance or product is brought to their attention. A medical office must accept the responsibility to obtain, evaluate, and act on any information pertaining to equipment hazards and the medication it may dispense.

Safe Medical Devices Act

In 1990 Congress passed the Safe Medical Devices Act (SMDA), which narrows the responsibilities to certain fundamental aspects that include:

- Medical offices or tangential services must report any incident of death, serious injury, or illness, if it is suspected that a medical device was the contributing factor.
- Manufacturers of devices that are permanent implants, and whose malfunction may pose a serious health problem or death, must conduct postmarked surveillance. Manufacturers are also required to establish methods for locating patients who depend on these devices. Because it is the medical facility that serves as a distributor for such devices, all medical offices must comply with reasonable requests from the manufacturer for additional information about the product.

- Under the SMDA, the FDA is authorized to institute recalls or stop-use notices on medical devices. If medical professionals and practices fail to do so, the FDA is authorized to impose civil penalties.

In order for the FDA to keep accurate records on the performance of drugs, as well as biological and medical devices, reports from manufacturers and user facilities are mandatory. However, reporting by health care professionals is voluntary.

MedWatch, the Medical Device Reporting Program

To further refine the process of medical products reporting, the FDA initiated MedWatch, the Medical Device Reporting (MDR) Program. MDR is the mechanism for the FDA to receive notice of significant medical device adverse events from manufacturers, importers, and user facilities so they can be detected and corrected quickly. MedWatch is an initiative designed both to educate all health professionals about the critical importance of being aware of, monitoring for, and reporting adverse events and problems to FDA or the manufacturer and to ensure that new safety information is rapidly communicated to the medical community, thereby improving patient care. This initiative provides the FDA with control over the safety and effectiveness of medical devices of all types, including diagnostics. MedWatch goals include:

- Increasing awareness of drug- and device-induced disease.
- Clarifying what should (and should not) be reported to the agency.
- Making it easier to report by operating a single system for health professionals to report adverse events and product problems to the agency.
- Providing regular feedback to the health care community about safety issues involving medical products.

Manufacturers of medical devices are required to register with the FDA and follow clearly outlined quality control procedures. Some of the medical devices must receive premarket approval from the FDA, while others must meet FDA performance standards.

MedWatch encourages health practitioners to promptly and voluntarily report any serious adverse health or product problems. Nurses, doctors, or other health care professionals are the earliest witnesses if a drug or device is not performing as anticipated.

Center for Devices and Radiological Health

Another program within FDA's jurisdiction is the Center for Devices and Radiological Health (CDRH). This program protects the public's health by providing reasonable assurance of the safety and effectiveness of medical devices and eliminating unnecessary human exposure to radiation emitted from electronic products. CDRH implements the SMDA and runs the MedWatch program. CDRH provides workshops and develops documents, such as information updates on breast implants.

The *FDA Desk Guide for Reporting Adverse Events and Product Problems* (DHHS Publication No. [FDA] 93-3204)[2] is a resource for additional information for voluntary reporting. It contains instructions for completing the MedWatch form and other important information on reporting, such as examples of the types of events to report, along with sample forms.

Mandatory reporting guides are also available from the FDA. Medical offices will benefit by obtaining copies of current guides and information, as there are significant changes regarding mandatory reporting.

US Food and Drug Administration, HFI-40, Rockville, MD 20857. Telephone 888 463-6332. Medical Device Office Telephone 800 332-1088. Website www.fda.gov.

US Department of Transportation

The DOT was established by an act of Congress in 1966. DOT develops and coordinates policies that include providing public transportation safety and maintaining the environment. The Research and Special Programs Administration (RSPA) operates the Office of Hazardous Materials Safety and provides services pertaining to safety, compliance, training, and research including protecting the public from the dangers inherent in the transportation of hazardous materials by air, rail, highway, and water. State hazardous waste transportation guidelines fall under the rules and regulations of the DOT.

US Department of Transportation, 400 7th Street, SW, Washington, DC 20201. Telephone 202 366-40000. Website www.dot.gov.

Drug Enforcement Administration

The mission of the DEA is to enforce the controlled substances laws and regulations of the United States. The Controlled Substances Act (CSA), Title II of the Comprehensive Drug Abuse Prevention and Control Act of 1970, is the legal foundation of the government's fight against abuse of drugs and other substances. This law is a consolidation of numerous laws regulating the manufacture and distribution of narcotics, stimulants, depressants, hallucinogens, anabolic steroids, and chemicals used in the illicit production of controlled substances.

Drug Enforcement Administration, US Department of Justice, 950 Pennsylvania Avenue, NW, Washington, DC 20530-0001. Telephone 202 305-8500. Website www.dea.gov.

Office of Justice Programs

Since 1984, the OJP has provided federal leadership in developing the nation's capacity to prevent and control crime. OJP implemented the 1994 Violence Against Women Act (VAWA) and led the national effort to stop domestic violence, sexual assault, and stalking of women. The VAWA of 2000 improves legal tools and programs that address domestic violence, sexual assault, and stalking. This act reauthorizes critical grant programs that were created by the 1994 act and subsequent legislation, establishes new programs, and strengthens federal laws.

Office of Justice Programs, US Department of Justice, 810 Seventh Street, NW, Washington, DC 20001. Telephone 202 307-5933. Website www.ojp.usdoj.gov. National Domestic Violence Hotline 800 799-SAFE (7233).

US Nuclear Regulatory Commission

The US Nuclear Regulatory Commission (NRC) oversees operation of nuclear power facilities and regulates the handling, use, and disposal of radioactive materials. The NRC is an independent agency established by the US Congress under the Energy Reorganization Act of 1974 to ensure adequate protection of the public health and safety, the common defense and security, and the environment in the use of nuclear materials in the United States. The NRC fulfills its responsibilities through a system of licensing and regulatory activities and adheres to Principles of Good Regulation, also known as the NRC Organizational Values, which are:

- *Independence.* Highest possible standards of ethical performance and professionalism.
- *Openness.* Nuclear regulation must be transacted publicly and candidly.
- *Efficiency.* Highest technical and managerial competence.
- *Clarity.* Coherent, logical, and practical regulations.
- *Reliability.* Based on best available knowledge from research and operational experience.

Radioactive waste is different from infectious and hazardous waste in that its danger cannot be treated to eliminate the threat. It cannot be put out like a fire, defused by submerging in water, cleansed with a chemical solution, or in any other way lessen or eliminate its harmful effects by human effort.

Only *time* can diminish the danger of radioactive waste. Over time, radioactive materials decay and become less hazardous. However, the rate of radioactive decay—called the half-life—can be very, very slow. Some materials actually require thousands of years to become thoroughly harmless.

In the interim, these substances emit dangerous radiation associated with cancer, birth defects, and other health problems. Consequently, these materials must be carefully handled during use and properly contained for disposal.

Neurologists and oncology clinics or labs may use radiological substances and are subject to the NCR's regulations regarding the types of materials that usually result in low-level radioactive waste.

Hazards and disposal techniques vary greatly among radiological materials and half-lives. In certain cases, contaminated materials are incinerated. Depending on their type and the applicable regulations, they may be left to decay in storage or disposed of through the sanitary sewer.

In an effort to make NRC documents and information readily available to the general public, NRC has placed its regulations, standards, and other documents on its Internet website.

US Nuclear Regulatory Commission, One White Flint North, 11555 Rockville Pike, Rockville, MD 20852-2738. Telephone 800 368-5642. Website www.nrc.gov.

Americans with Disabilities Act

To help persons with disabilities live independently and become economically self-sufficient, Congress passed the Americans with Disabilities Act (ADA) in July 1990. This act provides comprehensive civil rights protection for millions of people with disabilities in the following areas:

- Employment (Title I)
- Public services and transportation (Title II)
- Public accommodations (Title III)
- Telecommunications (Title IV)
- Other provisions (Title V)

The five federal agencies that are responsible for enforcing and regulating the ADA are:

- DOJ (public services and accommodations)
- DOT (transportation)
- EEOC (employment)
- ATBCB (architectural standards)
- FCC (telecommunications)

The ADA prohibits prospective employers from discriminating against persons with disabilities. According to the act, a qualified individual with a disability is one who satisfies the requisite skill, experience, education, and other job requirements and can perform the essential functions of the position, with or without reasonable accommodation.

This Act also states that any business or industry hiring a person with a disability must make reasonable structural or other process changes to allow the individual to perform the essential job functions. Wheelchair accessibility, such as ramps and wider doorways, are examples of structural changes. The Act further provides that these alterations not cause undue hardship or expense to the employer.

For medical facilities in modern buildings, most ADA-required structural changes have been corrected. However, for medical practices in converted houses or older office buildings, attention must be paid to ensure compliance with ADA requirements.

Under the ADA, job descriptions must clearly define the actual physical and mental requirements needed to perform the job. For instance, a written job description that states "lifting required" is insufficient information. The maximum amount of weight that may need to be lifted should be provided with the description.

If an applicant is denied employment because the prospective employer or designated hiring authority has decided the applicant cannot perform the necessary task or tasks, the job description becomes a critical issue in case of legal action against the employer. Job descriptions require analysis of the minimum physical capabilities and mental acumen required to successfully and safely perform the task. A job description must clearly explain what steps are needed to perform the job's function (eg, physical positioning, repetitions, torso movements). Trained professionals should perform the job function analysis, which may require the services of a consultant.

It is essential that medical practices have written policies and procedures that comply with the ADA guidelines, which not only protect the rights of persons with disabilities, but avoids preventable legal action as well.

US Department of Justice, 950 Pennsylvania Avenue, NW, Washington, DC 20530-0001. Telephone 800 514-0301. Website www.usdoj.gov.

OTHER FEDERAL AGENCIES

While some federal agencies do not have the authority to create or enforce the law, keeping abreast of trends or developments can help keep those in charge of safety and health in compliance and avoid unnecessary problems. This section describes federal agencies whose guidelines and recommendations can affect a medical practice. The agencies should be monitored periodically for new information.

Centers for Disease Control and Prevention

Agencies such as OSHA, EPA, and the Joint Commission on Accreditation of Healthcare Organizations (JCAHO) routinely rely on the research results and published recommendations and guidelines from the Centers for Disease Control and Prevention (CDC). The CDC is an agency of the Department of Health and Human Services that exists to promote health and quality of life by preventing and controlling disease, injury, and disability.

Through its weekly scientific publication, *Morbidity and Mortality Weekly Report (MMWR)*, medical practices can stay informed about data on specific health and safety topics. These include subjects such as controlling infection, dealing with infectious waste, or employee/patient protection from bloodborne infectious diseases, such as hepatitis B or acquired immunodeficiency syndrome (AIDS).

Because of modern transportation and a far more mobile society, communicable diseases are more easily contracted, thus making the research conducted by the CDC invaluable. In its attempt to battle disease, the CDC has an increasing number of centers throughout the nation. Listed below are some of the agencies and their specific purposes:

- *The National Center for Injury Prevention and Control.* Works to find prevention for injuries resulting from nonoccupational causes such as unintentional or violent acts.
- *The National Center for Infectious Diseases.* Seeks ways to prevent infectious diseases.
- *The National Center for Health Statistics.* Keeps track of the health of Americans.
- *The National Center for Chronic Disease Prevention and Health Promotion.* Seeks ways to prevent death and disability from chronic diseases.

- *The National Center for Environmental Health.* Works to prevent disability or death from environmental causes.
- *The National Institute for Occupational Safety and Health (NIOSH).* Conducts scientific research and makes recommendations for the prevention of work-related disease and injury.
- *The National Center for Chronic Disease Prevention and Health Promotion.* Assesses and reports on chronic diseases and their risk factors.
- *The National Center for HIV, STD, and TB Prevention.* Provides information, reports, counseling, and surveys.

Centers for Disease Control and Prevention, 1600 Clifton Road, Atlanta, GA 30333. Telephone 404 639-3311. Website www.cdc.gov.

National Institute for Occupational Safety and Health

NIOSH, an agency of the CDC, is the only government agency funded to evaluate scientific literature, direct studies, and provide scientifically valid recommendations to prevent worker exposure to hazards. It is responsible for conducting research on the full scope of occupational disease and injury ranging from lung disease in miners to carpal tunnel syndrome in computer users. In addition to conducting research, NIOSH:

- Investigates potentially hazardous working conditions when requested by employers or employees
- Makes recommendations and disseminates information on preventing workplace disease, injury, and disability
- Provides training to occupational safety and health professionals

The Occupational Safety and Health Act of 1970, which established OSHA, also established NIOSH. Although NIOSH and OSHA were created by the same Act of Congress, they are two distinct agencies with separate responsibilities. Congress decided that OSHA and NIOSH should be in separate cabinet level departments to avoid possible conflicts of interest. They believed that NIOSH could effectively conduct its mandated scientific research based on the overall health and safety of the American worker and not implementation issues. OSHA, however, must find the means to incorporate NIOSH's research and data into mandated regulations. OSHA must also take into consideration economical and technical feasibility factors.

NIOSH is in the US Department of Health and Human Services and is a research agency. OSHA is in the US Department of Labor and is responsible for creating and enforcing workplace safety and health regulations. NIOSH and OSHA work together toward the common goal of providing a safe work environment and protecting the health of each worker.

National Institute for Occupational Safety and Health, Hubert H. Humphrey Building, 200 Independence Avenue, SW, Room 7154, Washington, DC 20201. Telephone 800 356-4674. Website www.cdc.gov/niosh.

STATE REGULATIONS

Individual states have requirements for medical offices, practices, or labs on many of the same topics as federal agency regulations. However, the state requirements may be more rigorous or extensive than federal standards. These state rules usually are enforced through licensure or the periodic inspection and relicensure process.

The following is a representational list of various areas in which individual states may regulate medical practices:

- *Occupational safety and health.* Federal OSHA gives states the option of implementing their own workplace safety and health programs instead of adopting the federal program. State OSHA programs must be at least as stringent as the federal program.
- *Hazardous waste.* States may establish their own hazardous waste management agencies to assist in compliance with EPA and DOT rules. Requirements include proper storage, disposal, preparation for shipment, and transportation of hazardous materials, as well as training employers to recognize proper identification and control of hazardous substances.
- *Infectious waste.* Although this practice is beginning to change, state agencies usually create rules for infectious waste. The board of health or the state EPA, if there is one, usually issues the regulations. In addition, local sewer districts may have rules for the disposal of hazardous or infectious materials through the sewer.
- *Workers' compensation.* A common denominator for most states is the existence of workers' compensation laws. The state's attorney general or an agency that handles similar matters usually enforces these laws. The penalties for failure to comply with these laws, as well as the actual compensation to injured employees, are cumulative and can cause financial hardship for any size facility. Effective accident prevention and comprehensive safety programs can minimize the high cost of workers' compensation.
- *Fire codes.* The state or local fire authority, usually the fire marshal's office, verifies that fire safety requirements are met using its own state codes, OSHA regulations, and NFPA codes.
- *Health codes.* Health department officials, or the authority having jurisdiction, verify compliance with required infection control techniques and sanitation practices.

Because state and local requirements on these and other subjects may differ from or expand upon national standards, it is crucial that medical practices and labs know the regulations that affect them. As with federal mandates, medical practices should be aware that regional or state requirements might change as new information becomes available. Abiding by state and local regulations can provide a strong foundation for compliance with federal and voluntary standards.

NONGOVERNMENTAL ORGANIZATIONS AND AGENCIES

There are many nongovernmental organizations and agencies that deal with safety and health issues. Some of the more widely known are the American Conference of Governmental Industrial Hygienists, Inc (ACGIH), the American Society of Heating, Refrigerating, and Air-Conditioning Engineers (ASHRAE), the JCAHO, the NFPA, the American National Standards Institute (ANSI), and the Compressed Gas Association (CGA).

American Conference of Governmental Industrial Hygienists, Inc

When writing the Occupational Safety and Health Act of 1970, Congress recognized that it would take some time for the standard-setting process to begin working. They allowed OSHA to adopt guidelines for air contaminant concentrations from lists produced by nongovernmental consensus organizations. Therefore, the first OSHA standards governing chemical exposures were the threshold limit values (TLVs) published by the American Conference of Governmental Industrial Hygienists (ACGIH) in 1960. The TLVs were renamed permissible exposure limits and can be found in OSHA regulations (Standard Number 29 Code of Federal Regulations [CFR] 1910.1000, Table Z-1000, Subpart Z – *Toxic and Hazardous Substances*). There are about 400 chemicals listed in Table Z-1000.

For more than 60 years, ACGIH has been an organization with a goal to encourage the interchange of experience among industrial hygiene workers and to collect and make information and data accessible to them. Undoubtedly the best known of ACGIH's activities, the Threshold Limit Values for Chemical Substances Committee, was established in 1941 as a subcommittee of the Committee on Technical Standards. This group was charged with investigating, recommending, and annually reviewing exposure limits for chemical substances. Today's list of TLVs includes more than 700 chemical substances and physical agents, as well as 50 biological exposure indices for selected chemicals.

ACGIH also provides publications and professional reference texts and supports educational activities that facilitate the exchange of ideas, information, and techniques.

American Conference of Governmental Industrial Hygienists, Inc, 1330 Kemper Meadow Drive, Suite 600, Cincinnati, OH 45240. Telephone 513 742-2020. Website www.acgih.org.

American Society of Heating, Refrigerating, and Air-Conditioning Engineers

With chapters throughout the world, American Society of Heating, Refrigerating, and Air-Conditioning Engineers (ASHRAE) is an organization of 50,000 members. The society's sole purpose is to advance the arts and sciences of heating, ventilation, air conditioning, and refrigeration for the public's benefit through research, standards writing, continuing education, and publications.

Through its membership, ASHRAE writes standards that set uniform methods of testing and rating equipment and establishes accepted practices for the HVAC&R industry worldwide, such as the design of energy-efficient buildings.

Contractors use the established standards and accepted practices of this organization when building new facilities or remodeling medical facilities to ensure that ventilation systems are both adequate and support the intended use of the facility.

American Society of Heating, Refrigerating, and Air-Conditioning Engineers (ASHRAE), 1791 Tullie Circle, NE, Atlanta, GA 30329. Telephone 800 527-4723. Website www.ashrae.org.

Joint Commission on Accreditation of Healthcare Organizations

As early as 1910, Ernest Codman, MD, proposed the "end result system of hospital standardization," which recommended that hospitals track every patient to determine if the treatment had been effective. If the patient did not respond favorably, hospitals should determine why so that future treatments would be more successful.

By 1951, regional and national health care associations appeared across the nation seeking a stronger, more committed voluntary compliance movement. Several of these groups, including the American Hospital Association (AHA) and the American Medical Association (AMA), founded the Joint Commission on Accreditation of Healthcare Organizations (JCAHO). Its mission statement summarizes JCAHO's goals:

> The mission of the Joint Commission on Accreditation of Healthcare Organizations is to continuously improve the safety and quality of care provided to the public through the provision of health care accreditation and related services that support performance improvement in health care organizations.

During the intervening years, JCAHO's involvement has been primarily with health care facilities, such as hospitals, nursing homes, and behavioral health organizations. The following is a list of reasons why health care organizations seek Joint Commission accreditation (as extracted from *About the JCAHO*[3]):

- Assists organizations in improving their quality of care
- May be used to meet certain Medicare certification requirements
- Enhances community confidence
- Provides a staff education tool
- Enhances medical staff recruitment
- Expedites third-party payment
- Often fulfills state licensure requirements
- May favorably influence liability insurance premiums
- Enhances access to managed care contracts
- May favorably influence bond ratings and access to financial markets

The Joint Commission established the Ambulatory Health Care accreditation program in 1975 to encourage quality patient care in all types of freestanding ambulatory care facilities. A wide variety of ambulatory care organizations seek accreditation including group medical practices, military clinics, multispecialty group practices, physician offices, and single specialty providers.

Any health care facility may apply to JCAHO for accreditation once certain basic requirements have been met. To earn and maintain accreditation, an organization must undergo an on-site survey by a Joint Commission survey team at least every 3 years (laboratories must be surveyed every 2 years).

In a recent document, the Joint Commission wrote, "Today's health care environment is changing rapidly, and ambulatory care providers are experiencing new competitive pressures in the health care marketplace. Providing high-quality care to patients and continually improving performance are benchmarks of success, but it is increasingly important to demonstrate quality of care to payers, regulatory agencies, and managed care organizations. A growing number of ambulatory care organizations seek Joint Commission accreditation because our standards represent a national consensus on high-quality patient care."[4]

Joint Commission on Accreditation of Healthcare Organizations, One Renaissance Boulevard, Oakbook Terrace, IL 60181. Telephone 630 792-5000. Website www.jcaho.org.

National Fire Protection Association

For more than a century, the National Fire Protection Association (NFPA) has been dedicated to protecting people and their property from the devastating effects of fire. In some way, virtually every building, process, service, design, and installation in society today is affected by codes and standards developed through NFPA's true consensus system. Through NFPA's *National Fire Codes*, as well as its education and community outreach programs, NFPA is a membership-based organization and is recognized throughout the world as the leading authoritative source of technical background, data, and consumer advice on fire problems, protection, and prevention. The federal government and many state and local governments accept the NFPA codes as the basis for fire prevention and construction.

Products include current and reliable information and are presented in the most effective and memorable way using technical authorities, education professionals, and creative artists. Fire Code seminars and conferences are offered by respected professionals and designed to improve job performance by increasing the understanding of Code requirements.

An important fire safety resource for medical offices and labs is the *NFPA 101 Life Safety Code*. It is the cornerstone of life safety in new and existing structures and has protected countless lives in the past 80 years. The code now addresses Ambulatory Health Care Occupancies as a separate section and establishes the codes for life safety by considering a number of elements, such as early-warning detection and alarm systems, exit identification and lighting, sprinkler systems, building care maintenance, and storage for all occupied buildings.

The *NFPA 101 Life Safety Code* also includes requirements for emergency preparedness plans and drills, exit arrangements, and portable fire extinguishers. The emergency preparedness section provides information necessary for the preparation and implementation of a medical office's individual plan.

Another source of overall information on the latest developments in fire protection systems, equipment, and techniques is the latest edition of the NFPA *Fire Protection Handbook*.

NFPA also works to make health care facilities safer for patients and staff by providing *NFPA 99: Health Care Facilities*. This is the nation's premier source for data on the life-threatening hazards of electrical systems used in health care facilities. It includes up-to-date information on vital changes concerning gas and vacuum systems, including waste anesthetic gas disposal systems and gas shutoff valves.

In any medical practice someone must be in charge of fire safety and must be aware of the latest fire safety codes and standards to effectively prevent and survive fire.

National Fire Protection Association, 1 Batterymarch Park, PO Box 9101, Quincy, MA 02269-9101. Telephone 617 770-3000. Website www.nfpa.org.

American National Standards Institute

The American National Standards Institute (ANSI) was founded more than 80 years ago by five engineering societies and three government agencies as a private, nonprofit, membership organization, serving as an "umbrella" for the US standardization community. At the time, agencies with similar responsibilities had overlapping and sometimes conflicting standards. ANSI's mission is to enhance the global competitiveness of US business and the American quality of life by promoting and facilitating voluntary consensus standards and conformity assessment systems while ensuring their integrity. Conformity assessment involves evaluating products, processes, or services to determine if they adhere to a set of specific requirements.

ANSI itself does not develop standards. Rather, technical societies, trade associations, and other groups are accredited by ANSI to coordinate and lead technical development efforts. Accredited developers voluntarily submit the result of their work as candidate standards to ANSI for approval as "American National Standards." The ANSI designation signifies that ANSI's criteria for due process have been met and that a consensus for approval exists among those persons who were directly and materially affected and who chose to participate in the approval process.

ANSI standards apply to all areas of industry including health care, such as the safe use of medical lasers and personal protective equipment (PPE).

American National Standards Institute, 1819 L Street, NW, 6th Floor, Washington, DC 20036. Telephone 202 293-8020. Website www.ansi.org.

Compressed Gas Association

Medical offices routinely use a number of compressed gases, including carbon dioxide, anesthetic gases, and oxygen. Because these gases are stored under tremendous pressure, the slightest disturbance can cause this pressure to release in a destructive or deadly manner causing explosion, fire, injury, and property damage.

Since 1913, the Compressed Gas Association (CGA) has been dedicated to the development and promotion of safety standards and safe practices in the industrial gas industry. It provides technical advice and safety coordination for businesses in these industries. The CGA is also concerned with the handling of compressed gases wherever they are used, including medical offices that require these gases to adequately treat their patients. The CGA's mission is to (extracted from *About CGA*[5]):

- Gather and disseminate information about incidents and accidents involving industrial gases to eliminate recurrence.
- Cooperate with federal, state, provincial, and local government departments and regulatory agencies in developing responsible regulations for the industry.
- Develop, publish, and promote technical information.
- Conduct seminars and produce training programs.
- Support and coordinate research and development.
- Work with other organizations to address mutual concerns.

Medical personnel must be apprised of the potential dangers of mishandling these compressed gases. The CGA offers technical publications and audiovisual materials to safeguard the staff, visitors, and premises. It also advises the NFPA in developing compressed gas standards.

Compressed Gas Association, 1726 Jefferson Davis Highway, Suite 1004, Arlington, VA 22202-4102. Telephone 703 412-0900. Website www.cganet.com.

GUIDELINES FOR REGULATORY INSPECTIONS

Generally, medical offices or practices are not subject to inspection unless a fatal or serious accident occurs. As a precaution, each practice should develop policies and procedures for dealing with an inspection.

Although many regulatory agencies give advance notice of their inspections, they are not required to do so. The agency least likely to give advance notice is OSHA. In fact, alerting an employer of a pending inspection by OSHA can lead to a criminal fine or a jail term.

Management should develop a call list containing the names and telephone extensions of key individuals to notify in case of inspection. A typical inspection involves the following four steps:

1. *Arrival.* A staff member should greet the inspectors as promptly as possible and make them comfortable. The member should request their identification, if not already visible, and, if responding to a specific complaint, copies of any warrants or other documents.

    In large buildings or medical complexes with many different medical practices, the staff member should determine if the inspectors are at the correct location. If not, the member should give the inspectors directions, if known.

    The responsible staff member should first ask if the inspection covers the entire medical practice or facility or if it is limited to a specific area (as might be the case with a complaint). This permits prompt notification of key individual(s) on the inspection call list.

2. *Opening Conference.* This is a brief meeting during which the inspectors state the reason for the inspection and estimated time of duration. Staff members in attendance should supply any information requested by the inspectors, but volunteer nothing else.
3. *The Inspection.* During a first-time inspection, the inspectors may not be familiar with the physical layout of the facility or with all of the services that are provided, so someone on staff should accompany them. It may save time and prevent the inspectors from becoming lost or from entering into a hazardous area not intended for viewing.

    If the same person cannot serve as a guide throughout the inspection, then ask someone in each department to assist. Make certain the inspectors are provided with necessary equipment and that all safety procedures are adhered to during the inspection. Be prepared to take notes of the exact areas being inspected, which employees were interviewed, and any comments made by the inspectors with regard to the inspection itself.

    Every effort should be made to replicate any documentation made by the inspectors. Also, if photographs are taken, photograph the same subjects. The time, location, and date should be noted for each picture. If the inspector has brought along a camcorder instead of a camera, it would probably be expedient to videotape the same areas.

    It is a good idea to practice a possible inspection in advance. Knowing what to anticipate and how to proceed can make an inspection a smooth, painless occurrence, while not knowing can create an atmosphere of needless anxiety and nervousness.

    As is true in any situation that may involve legalities (eg, violations or infractions that might necessitate legal proceedings), the staff must understand never to volunteer information. Being prepared, even with short- or no-notice, can facilitate matters. Preparedness includes:

    - Knowing where all PPE is kept or stored
    - Being informed about which PPE is correct for specific tasks
    - Knowing where the keys to locked areas are kept
    - Informing staff where the camera, flash, and fresh film are kept

    If all staff members are cooperative, friendly, and prepared to accompany an inspector through a smooth inspection, then the integrity of the facility can only be enhanced.

4. *The Closing Conference.* Usually, at the end of an inspection, the inspector requests a closing conference. The person or persons responsible for the health and safety of the facility should be present. If there are unanswered questions, those best able to clarify the matter should attend the closing conference. As the conference draws to an end, ask the inspectors for a copy of their report. Based on the information in that report, prepare an in-house closing report detailing anything that requires improvement. This report should also include notes that were taken, photographs, and the names of employees who were interviewed. If the inspectors made any comments during the inspection that may prove useful in the improvement of the medical practice, these comments should be part of the

OTHER INFORMATION AND ASSISTANCE

The voluntary standards and government regulations affecting medical offices, laboratories, and other health care facilities are numerous and diverse. Individual practices must determine which regulations and guidelines apply to their particular services.

Compliance agencies and support organizations encourage individuals to take advantage of the information and assistance they provide. Many regulatory officials realize their standards may seem confusing and work to increase understanding through their publications and training. Many agencies also operate hot lines and provide Internet documents on specific topics. Medical practices that use the many resources available often find the regulations less intimidating and compliance an achievable goal.

All of these agencies and organizations (research, voluntary, or government compliance) exist to protect the health and safety of people in the workforce or those whom they serve. Table 2.1 lists the agencies discussed in this chapter and the scope of their services. Even if a citation or penalty is imposed on a medical practice, it could lead to improved conditions that avoid injury, illness, or even death.

TABLE 2.1
Summary of Agencies and their Scope

Governmental Compliance Agencies	Abbreviation	Scope
Occupational Safety and Health Administration	OSHA	Workplace safety and health
US Environmental Protection Agency	EPA	Control of the release of harmful materials into the environment
US Food and Drug Administration	FDA	Supervision of the development, testing, and monitoring of food, drugs, and medical devices
US Department of Transportation	DOT	Develops and coordinates policies that include providing public transportation safety and maintaining the environment
Drug Enforcement Administration	DEA	Enforces the controlled substances laws and regulations of the United States
Office of Justice Programs	OJP	Provides federal leadership in developing the nation's capacity to prevent and control crime
US Nuclear Regulatory Commission	NRC	The handling, use, and disposal of radiological materials
Americans with Disabilities Act	ADA	Help protect civil rights of the disabled under five federal categories
Other Federal Agencies		
Center for Disease Control and Prevention	CDC	Research, prevention, and control of disease
National Institute for Occupational Safety and Health	NIOSH	Evaluate scientific data, direct studies, and provide valid scientific recommendations to prevent worker exposure to hazards

continued

TABLE 2.1
concluded

Nongovernment Organizations and Agencies

American Conference of Governmental Industrial Hygienists, Inc	ACGIH	Collect, investigate, recommend, and review air contaminants, including threshold limit values for chemical exposures
American Society of Heating, Refrigerating, and Air-Conditioning Engineers	ASHRAE	International society to advance the arts and sciences of heating, ventilation, air conditioning, and refrigeration via research, standards writing, continuing education, and publications
Joint Commission on Accreditation of Healthcare Organizations	JCAHO	Primarily involved with health care facilities (eg, hospitals, nursing homes) to establish standards for safety and quality; provides accreditation to qualified facilities
National Fire Protection Association	NFPA	Reduce worldwide burden of fire and other hazards; develop scientifically based codes, standards, research, training, and education
American National Standards Institute	ANSI	Work to provide continuity and conformity standards to determine if they adhere to specific requirements
Compressed Gas Association	CGA	Develop and promote safety standards wherever compressed gases are used; advises NFPA in developing standards

State Regulations

State and local agencies		Workers' compensation, public health, civil rights protection for persons with disabilities; many areas also covered by federal agencies

ENDNOTES

1. Ramazzini B. *Diseases of Workers*. Translated from the Latin text *DeMorbis Articum* (1713) by Wilmer Cave Wright. OH&S Press; 1993. NorthWest Training and Development, Thunder Bay, Canada.
2. US Food and Drug Administration. *FDA Desk Guide for Reporting Adverse Events and Product Problems*. DHHS Publication No. (FDA) 93-3204. Available at: www.fda.gov.
3. Joint Commission on Accreditation of Healthcare Organizations. *About the Joint Commission on Accreditation of Healthcare Organizations*. Available at: www.jcaho.org.
4. Joint Commission on Accreditation of Healthcare Organizations. *Ambulatory Care Accreditation*. Available at: www.jcaho.org.
5. Compressed Gas Association. *About CGA*. Available at: www.cganet.com.

chapter 3

Infection Control

A Tale of Two Cities
It was the best of times, it was the worst of times, it was the age of wisdom, it was the age of foolishness, it was the epoch of belief, it was the epoch of incredulity, it was the season of Light, it was the season of Darkness, it was the spring of hope, it was the winter of despair, it was the era of people not washing their hands after using the bathroom, it was the era of people eating with their hands and falling violently ill after transferring bacteria to each other—in short, it was not a very sanitary period.
Allegheny County Health Department[1]

Perhaps at no time in history has the world's population been more secure and more at risk from infectious disease than the present. The twentieth century has witnessed both the eradication of diseases, such as small pox and polio, and an increase in emerging infectious diseases, such as AIDS.

Even though vaccines are powerful weapons in the fight against tuberculosis, some standard arsenals of antibiotics are not effective because the responsible organisms have evolved into resistant strains. The emergence of AIDS and the resurfacing of tuberculosis are reminders that infectious diseases have not disappeared.

Even as advances in the understanding of genetics promise an end to many diseases and new ways to combat infections, they also pose the threat of unparalleled suffering from genetically engineered organisms in the hands of terrorists.

For a number of reasons, infectious disease control is now more difficult than ever before. For example, the ease of global travel has greatly facilitated the transmission of infectious diseases. Furthermore, vaccine-induced antibodies found in childhood vaccines gradually decline with time as the children grow into adulthood.

Nevertheless, one of the greatest health achievements throughout the world has been progress against infectious diseases. One hundred years ago infectious diseases were the leading cause of death, even in developed countries. While mortality rates from infections have declined dramatically, certain infectious diseases still cause serious illness and death and remain an important health problem. Although some dangerous diseases appear controlled, they can come back aggressively without adequate attention.

To quote Charles Dickens, it is indeed "the best of times and the worst of times" with respect to infectious diseases. In short, it is a time to reinforce the need for basic sanitation and to strengthen efforts to control infectious disease.

EXAMPLES OF INFECTIOUS DISEASES

Almost any transmissible infection may occur in the community at large or within a medical practice and can affect both employees and patients. Some infectious diseases that are known to be transmitted in health care settings include bloodborne pathogens, acute gastrointestinal infections, hepatitis A, herpes simplex, measles, mumps, rabies, rubella, viral respiratory infections, and tuberculosis.

The Center for Disease Control and Prevention (CDC) and the Occupational Safety and Health Administration (OSHA) infection control precautions and guidelines provide methods that can be used to reduce the transmission of infections from patients to health care personnel and vice versa. Bloodborne pathogens and tuberculosis are considered serious risks for employees and patients and are discussed in this chapter by drawing on CDC and OSHA guidelines and recommendations.

INFECTIOUS DISEASE TRANSMISSION

A variety of microorganisms, including bacteria, viruses, rickettsia, and spirochetes cause infectious diseases. In some cases, it is necessary to determine the genus and species of the organism before beginning effective treatment. In other cases, identification is difficult and may take a fair amount of time before a proper diagnosis can be made.

The most common means for the transmission of infectious disease include:

- Aerosolization of infected body fluids
- Inhalation of bacteria, bacterial spores, or virus particles
- Contact with contaminated body fluids, food, water, or surfaces
- Zoonotic transmission (ie, bites from infected animals, insects, and other organisms)

Aerosolization

Droplets expelled during coughing, sneezing, or talking may contain bacteria or virus particles. While not restricted to this method of transmission, some diseases that are easily spread through aerosolization include tuberculosis, influenza, diphtheria, pertussis, and the common cold.

Inhalation

Bacterial spores can remain airborne for a significant period of time before encountering a host to infect. To a large extent, weather conditions affect the spread of those spores—the cause of regional outbreaks. Once an infection actually occurs, other modes of transmission of the bacterial spore may come into play. Air travel can also facilitate the spread of bacteria, their spores, and virus particles through the practice of recirculating cabin air.

In modern office buildings, improperly maintained HVAC systems provide a safe harbor and vehicle for microorganism transmission.

Contact with Contaminated Material

Microorganisms thrive in most body fluids. For this reason, every effort must be made to keep potentially contaminated surfaces clean and disinfected, including:

- Employee and patient restrooms
- Patient examination rooms
- Patient waiting areas
- Employee break and lunchroom facilities, especially refrigerators and food preparation areas
- Cleaning stethoscopes between patients
- Washing hands between patients

Zoonotic Transmission

Many insects, mammals, and other animals carry infectious organisms that can be passed on to humans. This transmission can occur through a bite or from ingesting an infected animal. Human rabies cases occur primarily from exposure to rabid animals.

DEFENSE MECHANISMS

Susceptibility to infections depends on many things, including, but not limited to, consistent nightly rest, exercise, and a well-balanced diet. A healthy person has a better chance of being able to resist an infection than a person who is in poor health with a compromised immune system.

Many vaccines are available that can prevent transmission or mitigate the effect of an infectious disease. Decisions about vaccines can be made by considering the likelihood of exposure to vaccine-preventable diseases and the potential consequences of not vaccinating, the type of contact with patients and their environment, and the characteristics of the patient population within the medical practice.

The basic premise of disease transmission is the same as it was 100 years ago—hand washing and good personal hygiene are the first and foremost methods of preventing the transmission of infectious diseases.

INFECTION CONTROL RESOURCES

OSHA, CDC, and the National Institute for Occupational Safety and Health (NIOSH) publish regulations and guidelines related to medical surveillance and the control and treatment of some types of infectious diseases. For example, OSHA has developed a regulation concerning the control of infectious disease carried by bloodborne pathogens. The CDC conducts routine disease surveillance, establishes guidelines for vaccines, and conducts research on well-known and emerging diseases. NIOSH has developed guidelines and conducted studies intended to protect workers from acquiring infectious diseases on the job.

Health care workers have found the Internet to be a useful tool to exchange information and to obtain the results of recent research conducted by government, academia, and health care organizations.

INFECTION CONTROL PROGRAM ELEMENTS

Employees are more likely to comply with an infection control program if they understand its rationale. Therefore, education is a principal element of an effective program. Because the risk of infection varies by job category, infection control education should be modified accordingly. [In addition, some employees may need specialized education on infection risks that are related to their employment and on preventive measures that reduce those risks.] The organization of the infection control program is usually influenced by the size of the practice. As outlined in the CDC's 1998, *Guideline for Infection Control in Health Care Personnel*[2], additional elements are necessary to attain the infection control goals of the medical practice and are summarized as follows:

- *Coordination among employees.* This process helps to ensure adequate surveillance of infections in employees and offers a provision of preventive services. Coordinating activities also helps to ensure that investigations of exposures and outbreaks are conducted efficiently and preventive measures are promptly implemented.

- *Medical evaluations.* Before hiring, this step can ensure that employees are not placed in jobs that would pose undue risk of infection to them, other employees, patients, or visitors. An important component of this step is a health inventory, which usually includes determining immunization status and obtaining histories of any conditions that might predispose personnel to acquiring or transmitting communicable diseases. Physical examinations can also screen for a variety of conditions and be used as a baseline for determining whether future diseases are work related.

- *Health and safety education.* Employees are more likely to comply with an infection control program if they understand its rationale. Clearly written policies, guidelines, and procedures ensure uniformity, efficiency, and effective coordination of activities. However, because the risk of infection varies by job category, infection control education should be modified accordingly. In addition, some personnel may need specialized education on infection risks related to their employment and on preventive measures that will reduce those risks. Further, educational materials need to be appropriate in content and vocabulary to the educational level, literacy, and language of the employee. The training should comply with existing federal, state, and local regulations regarding requirements for employee education and training. All health care employees need to be educated about the organization's infection control policies and procedures.

- *Immunization programs.* Optimal use of vaccines can prevent the transmission of vaccine-preventable diseases and eliminate unnecessary work restriction. Prevention of illness through comprehensive immunization programs is far more cost-effective than case management and outbreak control. By implementing screening programs combined with tracking systems in the medical practice, more accurate records are likely to be maintained.

- *Management of job-related illnesses and exposures.* Measures must be implemented to ensure prompt diagnosis and management and to provide appropriate postexposure prophylaxis after job-related exposures. Further transmission of infection must be prevented, which sometimes warrants exclusion of employees from work or patient contact. Decisions on work restrictions should be based on the mode of transmission and the epidemiology of the disease. Policies should be designed to encourage employees to report their illnesses or exposures and not to penalize them with loss of wages, benefits, or job status.

- *Health counseling.* This process offers employees individualized information regarding the risk and prevention of occupationally acquired infections, the risk of illness or other adverse outcome after exposures, management of exposures, and the potential consequences of exposures or communicable diseases for family members, patients, or other employees.
- *Maintenance of records, data management, and confidentiality.* Effective recordkeeping helps to ensure that the medical practice provides consistent and appropriate services to employees. Individual records should be maintained in accordance with the OSHA Medical Records Standard and the OSHA Bloodborne Pathogens (BBP) Standard.

NEEDLESTICKS AND OTHER SHARPS INJURIES

Half of all health care workers are employed in nonhospital settings, such as outpatient and physicians' offices. The CDC estimates that 600,000 to 800,000 needlestick and other percutaneous injuries involving needles or other sharps contaminated with bloodborne pathogens occur among health care employees each year.

The International Health Care Worker Safety Center[3] was established in 1992 for the prevention of occupational transmission of bloodborne pathogens. This center provides a standardized system for recording bloodborne pathogen exposures, including software for entering, accessing, and analyzing the data—the Exposure Prevention Information Network (EPINet). The center also conducts epidemiological research and provides technical assistance to health care institutions. EPINet data from 1993 to 1998, comparing exposure risks and characteristics of injuries in hospital vs. office workers, concludes that employees in physicians' offices and outpatient settings have a proportionately higher frequency of injuries from blood-filled needles (those most likely to transmit bloodborne pathogens).

Preventing occupational exposures to bloodborne pathogens means making a commitment to a needlestick prevention program. Key areas of the program include blood-drawing, injection equipment, body fluid contact, waste disposal, training, and postexposure follow-up.

The preponderance of needlesticks occurs during recapping of the needle after use, yet before discarding. Recapping is inappropriate handling of nonreusable sharps and needles, unless there is no alternative. If recapping is essential because of a particular procedure, do so with a mechanical device or in a way that requires only one hand. For example, a scooping technique requires the employee to place the cap on its side, then insert the needle into it.

Contaminated sharps must not be sheared or broken. Sharps can be eliminated in many cases, such as switching from glass to plastic blood collection vacuum tubes, changing from glass to plastic microbore capillary tubes for measuring hermatocrit, or using alternative methods of measuring hematocrit that do not require capillary tubes.

Some sharps are meant to be used more than once (eg, scalpels, large-bore needles, saws). Because reusable sharps pose the same percutaneous exposure hazard as disposable sharps, contain them in a manner that eliminates or minimizes the hazard until they are reprocessed. Therefore, the containers for reusable sharps must meet the same requirements as containers for disposable sharps, with the exception that they are not required to be sealable. However, they must be designed to minimize pricks or cuts when transported for sterilization.

Improved engineering controls are among the most effective approaches to reducing occupational hazards and are an important element of a needlestick prevention program. Such controls include eliminating the unnecessary use of needles and implementing devices with safety features. Characteristics of safety devices include the following:

- The device is needleless.
- The safety feature is an integral part of the device.
- The device preferably works passively (eg, it requires no activation by the user). If user activation is necessary, the safety feature can be engaged with a single-handed technique that allows the employee's hands to remain behind the exposed sharp.
- The user can easily tell whether the safety feature is activated.
- The safety feature cannot be deactivated and remains protective through disposal.
- The device performs reliably.
- The device is easy to use and practical.
- The device is safe and effective for patient care.

Although each of these characteristics is desirable, some are not feasible, applicable, or available for certain health care situations. For example, needles are necessary when alternatives for skin penetration are not available. Also, a safety feature that requires activation by the user might be preferable to one that is passive in some cases. Each device must be considered on its own merit and its ability to reduce workplace exposures and injuries.

Whenever using sharp objects, concentrate on the task at hand. Many accidental exposures to contaminants occur because of internal or external distractions or failure to follow established procedures. Figure 3.1 lists recommendations to prevent sharps injuries.

After use, place sharps in a puncture-resistant container that is located as close as possible to where they will be used and in every area where one might be used. The staff, whether janitorial or nursing, must monitor these containers to be sure they are not overfilled and know how to dispose of them safely when at capacity. Allowing a disposal container to become too full is another leading cause of needlestick injuries.

FIGURE 3.1

NIOSH Recommendations to Prevent Sharps Injuries

Adapted with permission from the National Institute for Occupational Safety and Health Alert DHHS (NIOSH) Publication No. 2000-108.

Preventing Needlestick Injuries in Health Care Settings

WARNING!

Health care workers who use or may be exposed to needles are at increased risk of needlestick injury. Such injuries can lead to serious or fatal infections with bloodborne pathogens such as hepatitis B virus, hepatitis C virus, or human immunodeficiency virus (HIV).

Employers of health care workers should implement the use of improved engineering controls to reduce needlestick injuries:

- Eliminate the use of needles where safe and effective alternatives are available.

- Implement the use of devices with safety features and evaluate their use to determine which are most effective and acceptable.

Needlestick injuries can best be reduced when the use of improved engineering controls is incorporated into a comprehensive program involving workers. Employers should implement the following program elements:

- Analyze needlestick and other sharps-related injuries in your workplace to identify hazards and injury trends.

- Set priorities and strategies for prevention by examining local and national information about risk factors for needlestick injuries and successful intervention efforts.

- Ensure that health care workers are properly trained in the safe use and disposal of needles.

- Modify work practices that pose a needlestick injury hazard to make them safer.

- Promote safety awareness in the work environment.

- Establish procedures for and encourage the reporting and timely follow-up of *all* needlestick and other sharps-related injuries.

- Evaluate the effectiveness of prevention efforts and provide feedback on performance.

FIGURE 3.1

continued

Health care workers should take the following steps to protect themselves and their fellow workers from needlestick injuries:

- Avoid the use of needles where safe and effective alternatives are available.

- Help your employer select and evaluate devices with safety features.

- Use devices with safety features provided by your employer.

- Avoid recapping needles.

- Plan for safe handling and disposal before beginning any procedure using needles.

- Dispose of used needles promptly in appropriate sharps disposal containers.

- Report all needlestick and other sharps-related injuries promptly to ensure that you receive appropriate follow-up care.

- Tell your employer about hazards from needles that you observe in your work environment.

- Participate in bloodborne pathogen training and follow recommended infection prevention practices, including hepatitis B vaccination.

Disposing of nonreusable sharps and needles without recapping or attempting to bend or break these items and placing them in disposal containers that block any possible insertion of fingers or hands reduces the chance of needlesticks. Disposal containers generally have a slot adequate for discarding used syringes or needles and are usually designed so the used sharps stack flat on top of each other to avoid needlesticks. New sharps disposal containers continue to enter the market via ever-changing product manufacturers and distributors.

There are specific types of containers for sealed disposal, and each container type offers details on container use and disposal. NIOSH also provides documentation and other resources on sharps disposal containers and preventing needlestick injuries. To alleviate the emotional anguish that is felt by employees who experience a needlestick, all measures should be taken to avoid this problem. In addition to the emotional issues surrounding needlesticks, the cost of investigating an incident is estimated at several hundred dollars and the costs of treating HIV or hepatitis infection are much greater.

Needlestick Safety and Prevention Act

Compliance with OSHA's Bloodborne Pathogens (BBP) Standard (Standard Number 29 CFR 1910.1030) significantly reduces the risk of employees contracting a bloodborne disease during work. Nevertheless, occupational exposure to bloodborne pathogens from accidental sharps injuries in health care settings continues to be a serious problem. For that reason, the federal government has taken increased action on needlestick prevention. In November 2000, the Needlestick Safety and Prevention

Act was signed into law. This act directs specific revisions to the BBP Standard, including clarifying the requirement for employers to select safer needle devices as they become available and involving employees in identifying and choosing the devices.

Since publication of the BBP Standard in 1991, there has been a substantial increase in the number and assortment of effective engineering controls available to employers. There is now a large body of research and data concerning the effectiveness of newer engineering controls, including safer medical devices.

Numerous studies have demonstrated that using safer medical devices can be extremely effective in reducing sharps injuries when they are part of an overall bloodborne pathogens program. The CDC estimates that 62% to 88% of sharps injuries can potentially be prevented by the use of safer medical devices.

Training and education in the use of safer medical devices and safer work practices are significant elements in the prevention of percutaneous exposure incidents. Staff involvement in the selection and evaluation of a device is also important in order to achieve a reduction in sharps injuries, particularly as new, safer devices are introduced into the workplace.

OSHA'S STANDARD FOR BLOODBORNE PATHOGENS

Health care infection control policies traditionally have been geared toward monitoring and controlling the spread of nosocomial infection among patients. In recent years, the advent of life-threatening diseases, such as hepatitis B virus (HBV), hepatitis C virus (HCV), human immunodeficiency virus (HIV), and multiple-drug-resistant tuberculosis (MDR-TB) has brought new urgency to protecting health care workers.

In 1991, OSHA published the BBP Standard to reduce the health risk to workers whose duties involve exposure to blood or other potentially infectious materials. The standard specifies detailed compliance requirements and defines body fluids as semen, vaginal secretions, cerebrospinal fluid, or other potentially infectious material (OPIM). This standard applies to employees in a medical practice who routinely handle or are exposed to bloodborne pathogens, such as HIV, which causes AIDS, and HBV. Hepatitis Non-A, Non-B, or other nonmicroorganisms, Cytomegalovirus, Toxoplasma, and the viruses that cause venereal diseases are included in the OSHA standard. All these pathogens may be transmitted through blood and OPIM's that cause disease in humans.

In January 2001, OSHA promulgated a revised BBP Standard that became effective April 2001. Figure 3.2 presents the components of this new standard. The revised standard conforms with the requirements of the Needlestick Safety and Prevention Act, which includes the following:

1. New and expanded definitions (Standard Number 29 CFR 1910.1030 (b), Defnitions):

    - Expanding the definition of engineering control to include "safer medical devices, such as sharps with engineered sharps injury protections and needleless systems." (The revised definitions do not reflect any new requirements being placed on employers with regard to protecting employees from sharps injuries, but are meant only to clarify the original standard and to reflect the development of new, safer medical devices since that time.)

- Adding the term *Sharps with Engineered Sharps Injury Protections* and defined as "a nonneedle sharp or a needle device used for withdrawing body fluids, accessing a vein or artery, or administering medications or other fluids, with a built-in safety feature or mechanism that effectively reduces the risk of an exposure incident."
- Adding the term *Needleless Systems* and defined as "a device that does not use needles for: (A) the collection of bodily fluids or withdrawal of body fluids after initial venous or arterial access is established; (B) the administration of medication or fluids; or (C) any other procedure involving the potential for occupational exposure to bloodborne pathogens due to percutaneous injuries from contaminated sharps."

2. New requirements to the Exposure Control Plan. Employers must review their exposure control plans annually to reflect changes in technology that will help eliminate or reduce exposure to bloodborne pathogens. That review must include documentation of the employer's consideration and implementation of appropriate commercially available and effective safer devices.

3. Additional recordkeeping requirements for employers who are required to keep records. The new requirements include maintaining a sharps injury log of percutaneous injuries detailing specific information. The log must be maintained in such a way as to ensure employee privacy and will contain, at minimum, the following information:

    - Type and brand of device involved in the incident, if known
    - Location of the incident
    - Description of the incident

4. A new section that requires documented, solicited input from nonmanagerial employees who are responsible for direct patient care and who have direct potential for injuries from contaminated sharps in the identification, evaluation, and selection of effective engineering and work practice controls. Examples of employees include those in different departments of the medical practice.

The following section is a summary of the scope and application of the 1991 BBP Standard. The elements of the standard are continuing requirements of the revised 2001 Standard.

Exposure Control Plan

Employers who have workers with occupational exposures to blood or OPIMs must establish a written Exposure Control Plan designed to eliminate or minimize exposure. The plan must be available to employees and contain at least the following elements:

- *An exposure determination.* This document lists all job classifications in which *all* employees in those job classifications have occupational exposure. It also lists job classifications in which *some* employees have

CHAPTER 3 Infection Control   43

FIGURE 3.2

Bloodborne Pathogens Standards Requirements

Reproduced with permission from Chaff & Co, 2001.

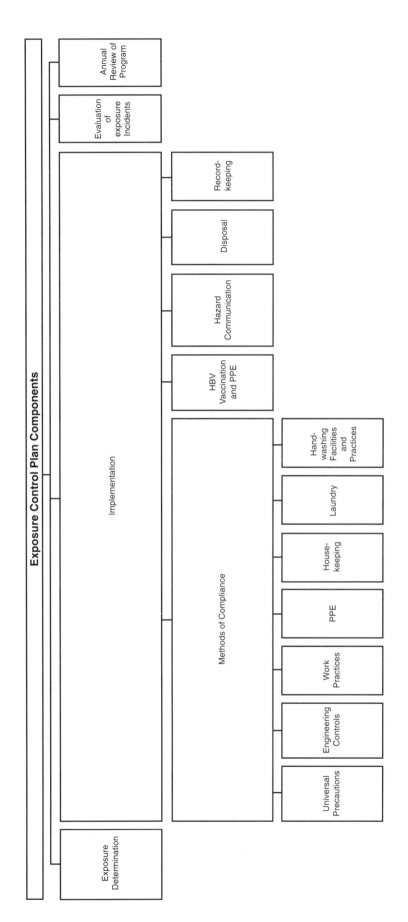

occupational exposure. (See Table 3.1 for a list of the exposure-prone jobs and procedures.)

- *The schedule and method of implementation of the components of the standard.* This element includes Methods of Compliance, hepatitis B Vaccination, Postexposure Evaluation and Follow-up, Communication of Hazards to Employees, and Recordkeeping.
- *The procedure for the evaluation of circumstances surrounding exposure incidents, as required by the standard.* Determines how an exposure occurred, such as improper use of equipment or a faulty procedure. Figure 3.3 provides an exposure chart defining the types of exposures likely to be encountered in a medical office.

Methods of Compliance

Methods of complying with the Exposure Control Plan include the following:

Universal precautions OSHA requires health care workers to observe universal precautions to prevent exposure to diseases that are transmitted via blood and other body substances. The OSHA requirement is based on the 1987 CDC document titled, *Recommendations for Prevention of HIV Transmission in Health-Care Settings.*[4] This document recommended that workers consistently use blood and body fluid precautions for all patients, regardless of their bloodborne infection status. Under universal precautions, blood and certain body fluids of all patients are considered potentially infectious for HIV, HBV, and other bloodborne pathogens. Figure 3.4 lists universal precautions for health care workers.

TABLE 3.1

Exposure-Prone Jobs and Procedures

Jobs

- Physician
- Nurse
- Nurse practitioner
- Housekeeping
- Laboratory technician

- Laundry
- Respiratory therapist
- Security officer
- Phlebotomist

Procedures

- Phlebotomy
- Intramuscular injection
- Intravenous access
- Emergency intobation
- Using power saws
- Suturing
- Handling contaminated laundry
- Assisting during surgery
- Collecting filled sharps containers
- Changing wound dressings

- Servicing contaminated equipment
- Shaving patients
- Pelvic exams
- Evaluating patients' oral hygiene
- Endotracheal suctioning
- Taking rectal temperatures
- Wound irrigation
- Specimen collection
- Endoscopic procedures

Adapted with permission from ERCI, *Special Report, Physician Office Safety Guide.* Copyright 1998.

FIGURE 3.3

Chart to Assess Type of Exposure
Reproduced with permission from Chaff & Co, 1993.

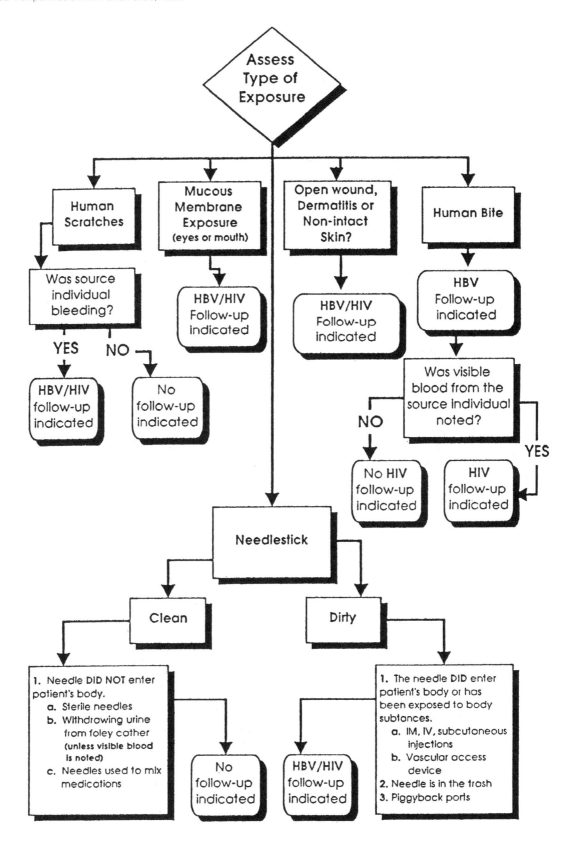

FIGURE 3.4
Universal Precautions for Heath Care Workers
Reproduced with permission from Chaff & Co, 1993.

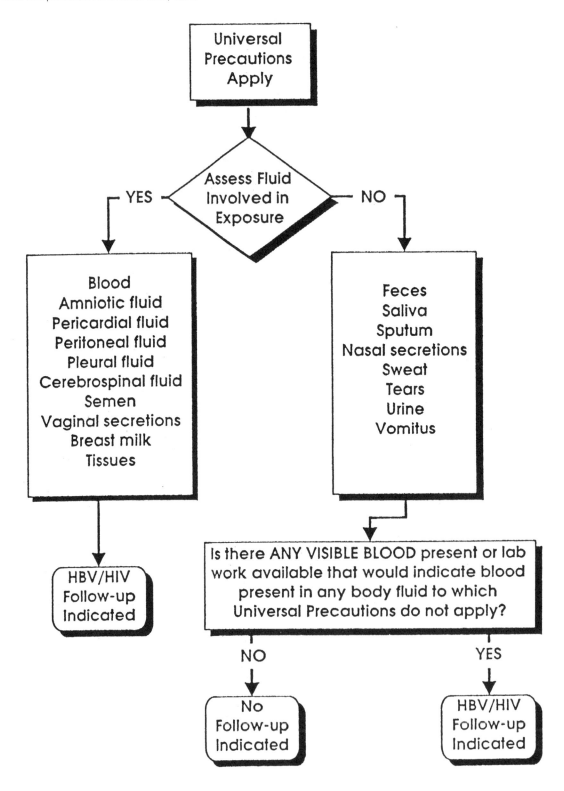

Engineering and work practice controls OSHA requires employers to institute engineering and work practice controls as the primary means of eliminating occupational exposure or reducing it to the lowest feasible extent.

1. **Engineering controls**. Engineering controls include all measures that isolate or remove a hazard from the workplace. OSHA encourages employers to involve employees in the selection of effective engineering controls to improve employee acceptance of the newer devices and to upgrade the quality of the selection process. Examples of such improvements include blunt suture needles and plastic or mylar-wrapped glass capillary tubes, as well as controls that are not medical devices, such as sharps disposal containers and biosafety cabinets.

   To eliminate or minimize exposures, employee training in using the engineering control is also required. Engineering controls must be examined, repaired, or replaced on a regular schedule to ensure they provide the intended protection. For safer devices, this step includes ensuring that the device is functioning effectively, protective shields are not removed or broken, and controls are functioning as intended.

2. **Work practice controls**. Following are examples of work practice controls:

   - *The no-hands procedures when handling contaminated sharps.* This procedure can refer to the use of self-closing and sealable, rigid sharps containers that prevent accidental contact and exposure after using the sharps. Employees do not need to handle the used sharps after disposal.
   - *The one-handed technique used to recap a needle.* The recapping of needles is not permitted unless using the one-handed technique or a mechanical device (eg, a pair of forceps) to place the cap back onto the used needle.
   - *Pipette suction bulbs.* Use this technique whenever transferring blood or body fluids from one receptacle to another.

Personal protective equipment (PPE) Where occupational exposure remains after implementation of engineering or work practice controls, PPE is required (eg, eye protection, gloves, masks) and employees must be trained to properly use the PPE.

OSHA requires employers to provide, at no cost to employees, appropriate PPE such as gloves, gowns, face shields or masks, eye protection, mouthpieces, resuscitation bags, pocket masks, or other ventilation devices. Wearing PPE not only protects the caregiver, but also the patient. PPE must prevent blood or OPIMs from passing through to or reaching the employee's work clothes, street clothes, undergarments, skin, eyes, mouth, or other mucous membranes under normal conditions of use and for the duration of time that the employee uses the protective equipment.

If laboratory coats or uniforms are intended to protect the employee's body from contamination, they are to be provided by the employer at no cost to the employee. Laboratory coats, uniforms, and the like that are used as PPE must be laundered by the employer and not sent home with the employee for cleaning. Scrubs are usually worn in a manner similar to street clothing and normally should be covered by appropriate gowns, aprons, or laboratory coats when splashes to skin or clothes are reasonably anticipated. A gown that is frequently ripped or falls apart under normal use would not be considered "appropriate PPE."

Resuscitator and emergency ventilation devices fall under the scope of PPE and must be provided by the employer for use in resuscitation. Employees must be trained in the types, proper use, their location, and so forth.

Housekeeping   Employers must ensure that the workplace is maintained in a clean and sanitary condition. Employers must determine and implement appropriate written schedules for cleaning and method of decontamination based upon the location within the facility, type of surface to be cleaned, type of soil present, and tasks or procedures performed in the area. Cleaning and decontamination must include the following:

- Work surfaces
- Protective coverings, such as plastic wrap, aluminum foil, or imperviously backed absorbent paper used to cover equipment and environmental surfaces
- Pails and cans intended for reuse
- Reusable sharps that are contaminated with blood or OPIMs

Laundry   Handle contaminated laundry as little as possible. Employees who have contact with contaminated laundry must wear PPE, such as gloves. Transport contaminated laundry in bags that are labeled or color-coded to allow recognition that they contain a biohazard.

Hand-washing facilities and practices   Employers must provide hand-washing facilities that are easily accessible to employees. Hand washing with soap and at least tepid running water must be performed as soon as feasible, particularly in cases of gross contamination, to adequately flush contaminated material from the skin. Alternative hand-washing methods can be used as an interim measure when soap and water are not a feasible means. In these cases, the employer must provide either antiseptic towelettes or an antiseptic hand cleanser along with clean cloth or paper towels until the employee can wash his or her hands with soap and running water.

HBV, Postexposure Program, and Follow-Up

The HBV series must be available to all employees who have occupational exposure, and postexposure evaluation and follow-up must be available to all employees who have had an exposure incident.

The CDC suggests that, when possible, employees be vaccinated against measles, mumps, tetanus, or any other disease likely to be encountered in a medical practice. However, different medical practices require different safeguards. Medical offices can establish which safeguards are appropriate by checking patient files to determine the patient population served by the practice.

Communication of Hazards to Employees

Warning labels must be attached to containers of regulated waste, refrigerators, coolers, specimen boxes, and freezers containing blood or OPIM. OSHA maintains specific requirements regarding types of labels and signs. Training requirements are also outlined in the BBP Standard.

Recordkeeping

Employers must establish and maintain an accurate record for each employee with occupational exposure to BBPs. The standard includes medical and training record requirements.

Summary of Measures to Reduce Risks of BBP Exposure

Education and training are key elements in reducing the risks of pathogen exposure, and employees must be adequately trained to carry out the requirements of the standard. The following summarizes OSHA's measures for reducing risks of contact with BBPs:

- *Good Hygiene Practices.* A commonsense approach to reducing the risks of pathogen exposure is to use good hygiene habits and practices. The single most important means of preventing and controlling the spread of infection is through proper hand washing and drying. Immediately and thoroughly wash and dry hands and any other skin surfaces if contaminated with blood or other body fluids; after removing gloves; or before eating, drinking, or smoking. Wash and dry hands before and after using the bathroom. Consume food and beverages in appropriate designated areas. Do not share personal items such as razors, lipstick, or other cosmetics. Avoid sharing the same food, beverages, tobacco products, and drinking containers.
- *HBV Protection.* Reducing the risk of infection from a BBP such as hepatitis B requires special protective measures. Because this virus is resistant to normal hygiene practices and can live outside the body for several days, there is an increased exposure potential. According to the CDC and OSHA, the most effective method of infection control against this particular pathogen is to be vaccinated before exposure.
- *Universal Precautions and Exposure Recognition.* Recognize work-related tasks and activities that may involve exposure to blood and other infectious materials, such as body fluids. Some jobs present a higher risk of exposure than others and may require additional protective measures. The type and extent of potential exposure should determine the amount of protection needed to perform various tasks. For example, wider ranges of personal protective clothing and equipment are needed in areas where there are exposure risks, such as splashes, sprays, splatters, or droplets of potentially infectious materials.
- *PPE.* Wear or use appropriate protective equipment to prevent exposure to skin and mucous membrane, if anticipating contact with blood or bodily fluids. PPE includes fluid-resistant gloves, gowns, aprons, face shields, respiratory equipment, and eye protection. PPE

must not permit blood or OPIMs to pass through to garments or reach skin, mouth, eyes, or other mucous membranes under normal work conditions. Do not take unnecessary chances.

- *Avoid Exposures.* To prevent injuries while handling or cleaning sharp instruments or disposing of used needles, give the task total attention. Distractions or interruptions can lead to improper handling and may ultimately lead to cuts, stabs, or pricks.
- *Housekeeping.* Determine and implement an appropriate written schedule of cleaning and decontamination based upon the location within the practice (eg, surgical room vs. patient examination room), type of surface to be cleaned (eg, hard-surfaced flooring vs. carpeting), type of soil present (eg, gross contamination vs. minor splattering), and tasks and procedures being performed (eg, laboratory analyses vs. routine patient care). The particular disinfectant used, as well as the frequency with which it is used, depends upon the circumstances in which the housekeeping task occurs. Decontaminate or dispose of contaminated materials, such as laundry, sharps, and PPE, in a safe, appropriate manner as described in the Housekeeping Section of the BBP Standard. "Appropriate disinfectants" are described in the standard. Avoid unnecessary handling of contaminated materials.
- *Disposal.* If an item cannot be effectively decontaminated, dispose of it in accordance with established regulations and procedures outlined in the standard. The standard defines the term "regulated waste" and outlines the categories of waste that require special handling. Properly contain and dispose of regulated waste to ensure it does not become of source of transmission of disease. OSHA specifies certain features of the regulated waste containers, including appropriate tagging. The ultimate disposal method (eg, landfilling, incinerating) for medical waste falls under the purview of the EPA and state and local regulations.
- *Identification of Labels and Signs.* Learn to recognize hazard warning signs, labels, color-coding systems, and containers used for regulated waste and other infectious materials. This knowledge can help prevent accidental mishandling and unnecessary exposure to bloodborne pathogens. Always exercise care in handling containers and/or materials displaying the biohazard label.

Dealing with an Exposure

If an exposure does occur on the job, the CDC recommends that:

1. Immediately following an exposure to blood:
   - Wash needlesticks and cuts with soap and water.
   - Flush splashes to the nose, mouth, or skin with water. (Splash irrigation should be with tempered water and should last 15 minutes.)
   - Irrigate eyes with clean water, saline, or sterile irrigants.

No scientific evidence shows that using antiseptics or squeezing the wound will reduce the risk of transmission of a BBP. Using a caustic agent, such as bleach, is not recommended.

2. Following blood exposure:

- Report the exposure to the department in the office responsible for managing exposures. Prompt reporting is essential because, in some cases, postexposure treatment may be recommended and should be started as soon as possible. Any exposure must be deemed as high risk and receive prompt treatment by a physician who will assess the degree of actual risk. Postponing examination and treatment could lead to dangerous consequences.

Refer to the CDC guidelines for additional direction for treatment and follow-up after exposure.

ALTERNATIVE INFECTION CONTROL METHODS

Body Substance Isolation (BSI) and standard precautions are alternative methods for infection control. The concepts incorporate not only the fluids and materials covered by OSHA's BBP Standard but expand coverage to include all body fluids and substances.

Body Substance Isolation

OSHA accepts BSI as an acceptable alternative to universal precautions provided the facilities using BSI adhere to all other provisions of the BBP Standard. BSI requires the following methods of control and prevention:

- Changing to clean gloves between patients and for contact activity
- Hand washing
- Additional PPE, as needed
- Sharps and needles placed in sharps containers
- Lab specimens handled as if they are infectious
- Handling and reprocessing practices for contaminated articles
- Soiled linen procedures
- Trash in plastic bags
- Private rooms with warning signs for patients with communicable diseases
- Other cautionary signs

Standard Precautions

The CDC recommends the use of an alternative system, or standard precautions, that incorporates universal precautions. The use of standard precautions, including appropriate hand washing and barrier precautions, further reduces contact with blood and body fluids.

The staff should take these precautions when potentially exposed to blood, all bodily fluids, all secretions and excretions other than perspiration, or if nonintact skin or mucous membranes are present and exposed. These situations are recognized as possible sources for contamination. Examples of other recognized possible sources are any unmarked vials or containers, if the contents are unknown. In these instances, there is no implication that there are pathogens or sources for infection—only that standard precautions are necessary to protect the staff from possible contamination.

OCCUPATIONAL EXPOSURE TO TUBERCULOSIS

Tuberculosis (TB) is a disease caused by a bacterial infection that is sometimes infectious. TB can affect many parts of the body, but is found most commonly (ie, 80% of the time) in the lungs, where it is called "pulmonary tuberculosis." The most common symptoms that could indicate TB are a cough that lasts longer than 3 weeks and having a fever for more than a week. Other symptoms include shortness of breath, weight loss, night sweats, loss of appetite, and sometimes lumps in the neck or swelling of joints. Physicians can perform skin tests to determine if someone has TB.

TB is a slowly developing disease, and the early stage might be passed off as a "cold" or "flu." If TB is not discovered early, the disease spreads. A cavity can form in the lung, which is an incubator of germs that can be spread by air droplets and threaten everybody who comes in contact with the sick person.

Active TB consumes the lungs, causing large lesions in the tissue. The large lesions prevent the lungs from working properly and can eventually lead to death in the untreated individual. Because of the infectious nature and usually long chronic course, TB affects the health and lives of many families. Until the 1800s, no treatment for TB existed.

Today, when TB infection is found, preventive treatment with drugs can often prevent further development of the disease. An ongoing and prevalent problem with attempts to control TB is that many people do not complete the prescribed course of treatment (ie, taking medication). A common reason for patients to discontinue medication is because they begin to feel significantly better in a matter of weeks and decide to stop the medication. Another reason is due to medication side effects, such as liver function abnormalities or gastrointestinal side effects, such as nausea, vomiting, and loss of appetite. A complete recovery occurs only if the medication is taken as prescribed.

The germs that cause TB are spread when people who have the disease cough, sneeze, laugh, or talk, affecting those around them. People breathe in the TB germs, which are droplets of *M. tuberculosis*, and they travel first to the small airways of the lungs. Exposure does not necessarily mean that an infection and symptoms will immediately follow. In some cases, the disease can remain dormant for years. If exposure to TB is known or suspected, a TB skin test is necessary for accurate diagnosis.

In 1998, a total of 18,371 active TB cases throughout the United States were reported to the CDC. An estimated 10 to 15 million people in this country have latent TB infections; about 10% of these will develop active TB at some point in their lives.

There are certain populations that are at greater risk of contracting TB than others. They include:

- Those infected with HIV or those with compromised immune systems.
- Those in close contact with active TB.
- Those who live in areas where TB is prevalent.
- People who travel to areas known to have a high incidence of TB.
- People who live in areas that are medically underserved.
- People who use intravenous drugs.
- Individuals with liver disease.
- Residents of communal facilities, such as long-term care, nursing homes, dormitories, correctional institutions, and homeless shelters.
- People 65 and older.
- Patients with medical conditions who are susceptible to tubercular bacteria.

Multiple-Drug-Resistant *Mycobacterium* Tuberculosis

TB is treated with antibiotics made to attack and kill the TB bacteria. In recent years, some strains of TB have become resistant to so many drugs that it is becoming difficult to obtain new drugs for treatment. These strains are known as Multiple-Drug-Resistant Mycobacterium Tuberculosis (MDR-TB). Antibiotic resistance is now a major public health issue. Even with treatment, this group of patients has a mortality rate of 40% to 60%—the same mortality rate as those who receive no treatment at all.

The cause of death from MDR-TB is frequently due to complications accompanying another primary disease, such as AIDS. MDR-TB is often found in groups of homeless people. This kind of TB also develops when people who are sick with TB do not take their drugs properly or for the prescribed period of time. When antibiotics are used incorrectly, bacteria can adapt and become resistant. Antibiotics are then no longer useful in fighting the bacteria.

Once bacteria have become resistant to a certain antibiotic, it no longer kills them. The resistant bacteria continues to grow when treated with the same drugs. Another drug to which they are not resistant must then be used. As a result, TB is treated with more than one type of drug at a time. Even if some of the bacteria in the patient are resistant to one or two of the drugs, the third antibiotic should make sure that the bacteria are killed.

Tuberculosis Infection Control Measures

The first step in preventing the spread of TB is to quickly identify, isolate and properly treat the contagious person. Nearly all TB patients under proper treatment will become non-contagious. Recommendations for preventing the transmission of TB for health care settings were originally established in CDC's Morbidity and Mortality Weekly Report (MMWR) entitled, *Guidelines for Preventing the Transmission of Tuberculosis in Health-Care Settings, with Special Focus on HIV-Related Issues* (1990/Vol. 39/No. RR-17).[5] In October of 1994, those guidelines were revised in CDC's MMWR entitled, *Guidelines for Preventing the Transmission of Mycobacterium Tuberculosis in Health-Care Facilities*, (1994/Vol. 43/No. RR-13)[6] and emphasized an effective TB control program that includes the following:

- Early identification, isolation, and treatment of persons with TB including a risk assessment
- Engineering controls, such as ventilation and TB isolation rooms
- Administrative procedures to reduce the risk of exposure, such as medical surveillance and employee education and training
- Respiratory protection that complies with OSHA requirements
- Decontamination processes, such as cleaning, disinfecting, and sterilizing patient-care equipment

The following is a summary of the recommendations pertaining to employee health as outlined in the CDC's, *Guidelines for Infection Control in Health Care Personnel*[7]:

1. **Strategies for prevention of transmission of TB.** Transmission of TB can be minimized by developing and implementing an effective TB control program based on a hierarchy of controls:
   a. Administrative controls
   b. Engineering controls
   c. Personal respiratory protection

   The risk of transmission of TB to or from employees varies according to the type and size of the medical practice, the prevalence of TB in the community, the patient population served by the practice, the occupational group that the person represents, the person's office area, and the effectiveness of the TB control program. A risk assessment is essential in identifying the nature of TB control measures that are appropriate for a particular medical practice, as well as for specific areas and occupational groups within that practice.

2. **TB screening program.** A TB screening program for employees is an integral part of a medical practice's comprehensive TB control program. Base the screening program on the practice's risk assessment. Baseline purified protein derivative (PPD) testing is an important element of the screening program.

3. **Follow-up evaluation.** The risk assessment indicates the employees who require annual testing and also evaluates those who have had previous positive PPD testing.

4. **Management of personnel after exposure to TB.** After recognizing TB exposure, administer PPD tests to employees as soon as possible. Such immediate PPD testing establishes a baseline to compare with

> **BOX 3.1**
>
> **Two-Step Tuberculin Skin-Test Screening**
>
> Some persons who are skin tested many years after infection with *M. tuberculosis* have a negative reaction to an initial tuberculin skin test, followed by a positive reaction to a subsequent skin test. This phenomenon is referred to as a boosted reaction. Boosted reactions to tuberculin skin tests are common in older adults and in persons who have been vaccinated with Bacillus of Calmette and Guérin (BCG). To reduce the likelihood of misinterpreting a boosted reaction as a new infection, two-step testing is used for baseline testing of persons who periodically will receive tuberculin skin tests. If the result of the first test is positive, the person is considered infected with *M. tuberculosis*; if negative, a second test should be administered 1 to 3 weeks later.
>
> A positive reaction to the second test probably represents a boosted reaction and is not considered a skin-test conversion. Persons who have a negative reaction to the second test should be classified as uninfected. In persons who have a negative skin-test result, a positive reaction to any subsequent test is considered a skin-test conversion and is likely to represent new infection with *M. tuberculosis* (DHHS, June 7, 1996[8]).
>
> Other examinations and monitoring may be necessary, which may be useful in guiding individual decisions regarding preventive therapy in selected situations.

subsequent PPD tests. Persons already known to have reactive PPD tests need not be retested, but should still be screened annually. Employees with evidence of new infection (ie, PPD-test conversions) need to be evaluated for active TB. If active TB is not diagnosed, consider preventive therapy.

5. **Preventive therapy.** For employees with positive PPD-test results who were probably exposed to drug-susceptible TB, preventive therapy with Isoniazid (INH) and coadministration of vitamin B6 or Pyrodoxine is suggested, unless there are contraindications to such therapy. Alternative preventive regimens have been proposed for persons who have positive PPD-test results after exposure to drug-resistant TB.

6. **Work restrictions.** Employees with active pulmonary or laryngeal TB may be highly infectious. Exclusion from duty is suggested until they are noninfectious. When employees are excluded from duty because of active TB, the practice should have documentation from the employees' health care providers that they are noninfectious before they are permitted to return to duty. The documentation needs to include evidence that the employee is receiving adequate therapy, the cough has resolved, and smears are negative. Periodic documentation must be provided from the health care provider regarding therapy and work restrictions.

7. **Considerations for BCG vaccine.** BCG vaccination has not been routinely used in the United States to protect health care personnel. Because of the resurgence of TB in the United States and new information about the protective effect of BCG, the role of BCG vaccination in the prevention and control of MDR-TB has been reevaluated.

Consultation with local and state health departments is advisable when determining whether to provide a BCG vaccination to employees.

Respirator Requirements

In February 1996, OSHA issued a directive to compliance officers outlining enforcement procedures for occupational exposure to TB (CPL 2.106).[9] In October 1997, OSHA published a proposed rule, *Occupational Exposure to Tuberculosis*, which referenced CDC's 1994 established guidelines.[10] The enforcement procedures require the use of respirators when employees are exposed to patients with TB. When using respiratory protection, the medical practice must develop and follow a written respiratory protection program. Only NIOSH-certified respirators can be used.[11]

The written respirator program must include the following elements:

- Clearly written and easily understood standards and procedures.
- Types of medical screening in effect (to be updated routinely).
- Training on the various types of respirators in use, including selection, handling, maintenance, and limitations.
- Training on how to check for correct fit of respirators, inclusive of fitting the face.
- Instruction on the proper care, cleaning, storage, and inspection of respirators.
- Routine inspection of all respirators to ensure they are in good working condition. If the respirator involves cartridges, these too must be inspected routinely. Cartridges can become defective and can result in serious injury, death, or property damage. The NIOSH certified N-95 respirator is the minimal respirator allowed to be used with TB patients.
- Scheduled and regular evaluations of the program, including reviews of new and available equipment.
- Regular reevaluations of the existing program to seek improved recommended procedures.

ENDNOTES

1. *The Literary Classics—A New Kind of Reading Material for Public Toilet Rooms.* 1997 Public Health Campaign. Allegheny County Health Department, Food Safety Program.
2. Bolyard EA, Tabian OC, Williams WW, Pearson ML, Shapiro CN, Deitchman SD, et al. *Guidelines for Infection Control in Health Care Personnel, 1998.* Special Article. Centers for Disease Control and Prevention, Public Health Service, US Department of Health and Human Services, Atlanta, GA.
3. International Health Care Worker Safety Center, Health Sciences Center, University of Virginia, Charlottesville, VA.

4. Center for Disease Control and Prevention. *Recommendations for Prevention of HIV Transmission in Health-Care Settings*. MMWR. August 21, 1987/Vol. 36/No. 2S.

5. Center for Disease Control and Prevention. *Guidelines for Preventing the Transmission of Tuberculosis in Health-Care Settings, with Special Focus on HIV-Related Issues*. MMWR. December 7, 1990/Vol. 39/No. RR-17.

6. Center for Disease Control and Prevention. *Guidelines for Preventing the Transmission of Mycobacterium Tuberculosis in Health-Care Facilities, 1994*. MMWR. October 28, 1994/Vol. 43/No. RR-13. Reprinted March 1994.

7. Bolyard EA, Tabian OC, Williams WW, Pearson ML, Shapo CN, Deitchman SD, et al. *Guidelines for Infection Control in Health Care Personnel*, 1998: 318–320.

8. Center for Disease Control and Prevention. *Prevention and Control of Tuberculosis in Correctional Facilities*. Recommendations of the Advisory Council for the Elimination of Tuberculosis. MMWR. June 7, 1996/Vol. 45/No. RR-8: 318-320. Reprinted March 1999.

9. Occupational Safety and Health Administration. *Enforcement Procedures and Scheduling for Occupational Exposure to Tuberculosis*. CPL 2.106. February 9, 1996.

10. Occupational Safety and Health Administration. *Occupational Exposure to Tuberculosis*; Proposed Rule – 62:54159-54309. October 17, 1997.

11. Occupational Safety and Health Administration. *Occupational Exposure to Tuberculosis*. Proposed Rule – 62:54159-54309, October 17, 1997.

chapter 4

Chemicals

By keeping employees well trained and informed and by enforcing good industrial hygiene practices, the chances of overexposure to hazardous chemicals will be minimized, and working conditions will be comfortable and safe. Susan L.P. Jordan, PhD, Union Carbide Corporation[1]

In today's society, it would be nearly impossible to live without the contribution of chemicals. In 1997, the US chemical industry produced about $389 billion worth of products and employed 1,032,000 workers. From toothpaste to shampoo, cookware to spacecraft, leading nations rely on the uses of chemicals to sustain and prosper in innumerable ways.

For years, chemicals have also been part of the everyday life in a medical practice. Whether it is applying rubbing alcohol to a patient prior to an injection or drawing blood to cleansing compounds or a variety of other uses within a practice—chemicals play an important role.

This poses a serious problem for exposed workers and their employers. Chemical exposure may cause or contribute to many serious health effects, such as heart ailments, kidney and lung damage, sterility, cancer, burns, and rashes. Some chemicals may also be safety hazards and have the potential to cause fires, explosions, and other serious accidents.

UNDERSTANDING OCCUPATIONAL EXPOSURE VALUES

There are several routes of entrance into the human body:

- Intact skin (absorption)
- Damaged skin
- The respiratory system (inhalation)
- The mouth (inhalation and ingestion)
- The eyes
- Accidental needle punctures

Not all substances can enter the body through all routes, although a chemical substance can enter through more than one. Figure 4.1 includes examples of routes of entrance into the body. Some substances can damage the skin or eyes directly without being absorbed, such as strong acid solutions, liquid nitrogen, or bleach. Other chemicals can damage the skin and be absorbed through the skin or by inhalation, such as many organic solvents and mercury. Still other chemicals, such as nitrous oxide, are an inhalation hazard.

FIGURE 4.1

Examples of Routes of Entry into the Body

Reproduced with permission from National Safety Council, *The Facts About The Hazard Communications Standard, A Guide Book.* Copyright 1985.

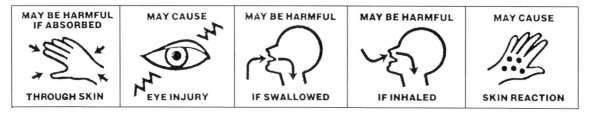

The exposure dose is the substance amount that actually enters the body during a specified time period. The substance remains in the body until it is metabolized or eliminated. Although some chemicals are rapidly metabolized into harmless substances or excreted unchanged, others are not and may be stored in the fatty tissues (eg, solvents), the lungs (eg, dusts, fibers), bone (eg, lead, radium), or blood (eg, soluble gases). Others may be metabolized into more toxic compounds.

For many chemicals, NIOSH's research and recommendations, as well as OSHA's regulations, call for a limit on the concentration of the substance in the air. These limits are referred to as a recommended exposure limit (REL-NIOSH) and a permissible exposure limit (PEL-OSHA). If the exposure limit is greater than permitted, OSHA requires the employer to switch to a less hazardous chemical or to install suitable engineering controls (eg, ventilation) to reduce exposure to less than the PEL. If needed, but only as a backup for engineering controls, OSHA requires employees to wear protective equipment (eg, respirators, protective eyewear) until achieving the OSHA standards.

Wherever chemicals are used, their concentration in the air usually fluctuates throughout the course of the day. This fluctuation is the result of variations in use and the nature of the process in which it is being used. In such cases, knowing the average concentration of the work shift is sufficient. For other chemicals, it is important to know the magnitude, duration, and frequency of the peak exposures. Yet with other substances, knowing both the average concentration throughout time as well as the peak concentrations is necessary.

Some of the frequently used abbreviations for occupational exposure values are:

- *C (Ceiling)*—The total, maximum concentration of exposure that must *not* be exceeded at any time during a work shift.
- *IDLH (Immediately Dangerous to Life and Health)*—The IDLH is based on a concentration that would affect the ability to escape the area within 30 minutes of exposure without the use of PPE. If a concentration could cause severe eye damage leading to blindness, it could be considered IDLH if the condition prevented the employee from escaping from the area. A situation could also be considered IDLH if exposure would result in any adverse and permanent damage to the employee's health during exposure and escape.
- *PEL (Permissible Exposure Limit)*—The time-weighted average exposure that must *not* be exceeded during an 8-hour work shift during a

40-hour workweek. PELs are OSHA values and are legally enforceable.
- *REL (Recommended Exposure Limit)*—The time-weighted average for up to a maximum 10-hour work shift during a 40-hour week, unless otherwise specified. These are NIOSH recommendations and are not legally enforceable.
- *PPM* (parts of substance per million parts of air).
- *STEL (Short-Term Exposure Limit)*—The mean concentration of exposure for a 15-minute work period that must *not* be exceeded during the work shift. The STEL time limit may vary depending upon the chemical, but is usually 15 minutes. STEL serves to prevent acute effects of exposure. Usually only four such periods are allowed in any 8-hour work shift. If the substance can cause both acute and chronic effects, the PEL may also be expressed as a combination of an 8-hour work shift and a 15-minute STEL. In general, no more than four such periods are allowed during any 8-hour work shift.
- *TLV (Threshold Limit Value)*—The time-weighted average for an 8-hour workday, unless otherwise noted. These are American Conference of Governmental Industrial Hygienists (ACGIH) values and, unless incorporated by reference by OSHA, are not legally enforceable.
- *TWA (time-weighted Average)*—The average concentration of exposure for a typical workweek (eg, 8-hour workday and a 40-hour week). This is the average concentration of a chemical substance in the air to which employees may be exposed. The purpose is to prevent adverse health effects of a chronic nature, such as lung or reproductive damage. TWA concentrations for airborne vapors are usually expressed as parts per million. Solid materials, such as dusts, are expressed as milligrams of substance per cubic meter of air.

CHEMICALS USED IN MEDICAL PRACTICES

Following is a listing of chemicals most often used in a medical practice:

Aerosolized Drugs	Glutaraldehyde
Ammonia	Hydrogen Peroxide
Carbon Dioxide	Iodine, Providone-iodine
Chemotherapeutic Drugs	Liquid Nitrogen
Chlorine	Mercury
Cleaning Products	Oxygen
Dilute Acids, Bases, and Stains	Phenolics
Formaldehyde	Nitrous Oxide

Through scientific studies, health hazard evaluations, criteria documents, current intelligence bulletins, and other publications, NIOSH provides guidelines to educate employees in medical practices about the hazards of chemicals, potential health effects, standards and recommendations, and exposure control methods. These extensive evaluations of the scientific literature include recommendations to OSHA for controlling exposures. OSHA and other organizations and agencies (eg, Agency for Toxic Substances and Disease Registry (ATSDR), EPA) also provide regulations, information, and recommendations for chemicals used in medical practices. Following is information and recommendations on the cited chemicals:

- Chemotherapeutic Drugs
- Formaldehyde
- Glutaraldehyde
- Mercury
- Phenolics

Chemotherapeutic Drugs

Nurses and pharmacists face a variety of potential hazards from contact with pharmaceuticals. The drugs of greatest concern are those associated with cytotoxicity and fetotoxicity (eg, folate antagonists, 6-mercaptopurine, some alkylating agents) and teratogenicity (eg, actinomycin-D, mitomycin-C, nitrogen mustard, prednisone, procarbazine, streptomycin, vincristine). Many chemotherapeutic agents have been reported to cause cancer in animals and thus can be considered to be potential human carcinogens (eg, cyclophosphamide, chlorambucil).

Antineoplastic drugs derive their names from the fact that they interfere with or prevent the growth and development of malignant cells and neoplasms. They may also be called cytotoxic or cytostatic because they have the ability to prevent the growth and proliferation of cells. Each year hundreds of thousands of cancer patients are treated with antineoplastic drugs.

Hazard Location
Cytotoxic drugs are prepared and administered in a wide variety of places, ranging from physicians' private consulting rooms to large pharmaceutical preparation rooms. At times, physicians or nurses may prepare cytotoxic drugs in inadequately ventilated patient-care or staff areas.

Characteristics (Physical Properties)
Cytotoxic drugs are toxic compounds and known to have carcinogenic, mutagenic, and/or teratogenic potential.

Potential Health Effects
When given to patients in therapeutic doses, many antineoplastic drugs are associated with an increased incidence of malignant tumors that develop at a later date.

Toxic effects have been observed in patients treated with antineoplastic drugs. These effects include lack of sperm production, reduced sperm counts, amenorrhea, as well as adverse effects on the bone marrow, heart, central nervous system, liver, skin, ears, pancreas, lungs, kidneys, and endocrine glands. Treatment with antineoplastic drugs has also resulted in depression of the hematopoietic system.

The acute effects of accidental exposure to these drugs can be severe. For example, an accidental needle prick of a patient's finger with mitomycin-C has been reported to cause the eventual loss of function of that hand.

Little is known about the potential health hazards of chronic exposure to antineoplastic drugs, but a study showed significant association between fetal loss and the occupational exposure of nurses to these drugs. It has been documented that nurses who handled antineoplastic drugs for a number of years develop liver damage. Nurses handling antineoplastic drugs have reported lightheadedness, dizziness, nausea, headache, skin and mucous membrane reactions, hair loss, cough, and possible allergic reactions. These side effects observed in nurses are the same as those noted by patients receiving antineoplastic drugs.

Methods for Estimating Exposure

Primary routes of worker exposure to antineoplastic drugs are inhalation and skin absorption. Exposures by inhalation can occur during drug preparation or administration. Aerosols can be generated when inserting needles into or withdrawing them from vials, and when expelling air from syringes before injection. Skin absorption may occur when antineoplastic drugs are spilled during their preparation or administration. Skin exposure may also occur as a result of contact with the urine of patients who are being treated with antineoplastic drugs.

Standards and Recommendations

There are currently no NIOSH RELs or OSHA PELs for antineoplastic drugs. Because of their properties, they must be regarded as highly toxic and handled according to regulatory requirements and guidelines.

Exposure Control Methods

Using a combination of specific containment equipment and certain work techniques, such as safe preparation and administration, nurses and physicians can effectively control the potential risks from repeated contact with parenteral cytotoxic drugs. For the most part, the techniques are merely an extension of good work practices by health care and ancillary personnel and similar in principle and practice to the use of universal precautions.

OSHA provides work practice guidelines for dealing with antineoplastic drugs. These guidelines address drug preparation, drug administration, waste disposal, spills, medical surveillance, storage and transport, training, and information dissemination.

Work practices Two elements are essential to ensure proper workplace practices:

- Education and training of all staff involved in handling any aspect of chemotherapeutic drugs
- A biological safety cabinet (BSC)

The costs of implementing the former and installing the latter are relatively minor. The potential benefits are major.

Personal Protective Equipment (PPE) Wear PPE when handling cytotoxic drugs, including drug preparation and caring for patients. Use appropriate gloves and change them regularly or immediately if torn or punctured. Wear protective disposable gowns and gloves while in the preparation area, but not outside the preparation area. A BSC is essential for the preparation of cytotoxic drugs. Where one is not currently available, an appropriate NIOSH-certified respirator provides the best protection until a BSC is installed. Surgical masks do not protect against the breathing of aerosols. Also, wear ANSI-approved splash goggles and a plastic face shield if a BSC is not in use and an eyewash fountain is not available.

Medical Monitoring
Workers exposed to antineoplastic drugs should receive preplacement and periodic medical evaluations that include at least the following:

- Complete work and medical histories.
- An examination that emphasizes the skin, liver, hematopoietic, reproductive, and nervous systems.

The examining physician has the discretion to perform other tests and should be particularly alert for symptoms of liver disease, skin and mucous membrane irritation, central nervous system depression, teratogenic effects, and cancer.

Formaldehyde

For many years, medical practices used formaldehyde as a disinfectant and currently use it as a preservative for tissues and biopsies. It is often combined with methanol and water to make Formalin.

Hazard Location
Formaldehyde may be used in the laboratory as a tissue preservative.

Characteristics (Physical Properties)
Pure formaldehyde is very reactive and polymerizes easily.

Potential Health Effects
Over time, it became evident to OSHA that formaldehyde could be highly injurious to the health of staff. Exposure of susceptible individuals to formaldehyde can cause allergic contact dermatitis. Not only is this colorless, pungent chemical harmful, but it is now a suspected carcinogen.

Acute effects The odor can be detected in air at about 0.8 ppm; however, even a short period of exposure will decrease the worker's ability to smell it. Therefore, odor is not a reliable warning for the presence of formaldehyde. Formalin solutions, if splashed in the eyes, may cause severe injury and corneal damage. Low ambient concentrations of formaldehyde (eg, 0.1 to 5 ppm) may cause burning and tearing of the eyes and irritation of the upper respiratory tract. Higher concentrations (eg, 10 to 20 ppm) may cause coughing, chest tightness, increased heart rate, and a sensation of pressure in the head. Exposures of 50 to 100 ppm may cause pulmonary edema, pneumonitis, and death.

Chronic effects Repeated exposure may cause some persons to become sensitized. Sensitization may occur days, weeks, or months after the first exposure. Sensitized individuals experience eye or upper respiratory irritation or an asthmatic reaction at levels of exposure that are too low to cause symptoms in most people. Reactions may be quite severe, with swelling, itching, wheezing, and chest tightness.

Dermatitis (including red, sore, cracking, and blistered skin) is also a common problem with formaldehyde exposure. Repeated exposure may make the fingernails soft and brown. A NIOSH health hazard evaluation[2] indicated that respiratory irritation, eye irritation, and dermatological problems were the primary health problems associated with formaldehyde exposure.

Formaldehyde is a mutagen in many assay systems and has caused nasal and other cancers in experimental animals. In 1981, NIOSH published the Current Intelligence Bulletin 34[3], which recommended that formaldehyde be handled as a potential carcinogen in the workplace.

Standards and Recommendations

OSHA developed the Formaldehyde Standard (Standard Number 29 CFR 1910.1048), which includes a surveillance program to identify and quantify the exposure levels of workers who were potentially exposed to formaldehyde. OSHA also developed the Laboratory Standard (Standard Number 29 CFR 1910.1450), which covers laboratory workers.

OSHA's PEL is 0.75 ppm of formaldehyde as an 8-hour TWA. This standard also includes a STEL of 2.0 ppm of formaldehyde in air during any 15-minute period. OSHA also established that, when an exposure reaches 0.5 parts of formaldehyde per million parts of air, certain measures must be put in place, such as determining the source of the exposure and instituting appropriate controls to reduce exposures below 0.5 ppm.

NIOSH's REL for formaldehyde is 0.1 ppm as determined in any 15-minute air sample and 0.016 ppm as an 8-hour TWA.

The ACGIH designated formaldehyde a suspected human carcinogen and recommended a TLV of 1 ppm (ie, 1.5 mg/m^3) as an 8-hour TWA with a STEL of 2 ppm (ie, 3 mg/m^3).

Environmental Monitoring

NIOSH industrial hygiene surveys have found formaldehyde concentrations ranging from 2.2 ppm to 7.9 ppm in medical facility autopsy rooms. All potential exposures to formaldehyde must be monitored. When monitoring results show work areas above the PEL, that area is established as a regulated area requiring special authorization to enter. When results are at or above the action level of 0.5 ppm, monitoring must be performed every 6 months.

Exposure Control Methods

If an appropriate substitute cannot be found, the next line of worker protection is the use of ventilation to remove vapors. In the event the ventilation is inadequate, employees must be furnished with and trained in the use of PPE and clothing, as well as the safe handling of the substance.

Engineering controls The following engineering controls are recommended to minimize formaldehyde exposure:

- Local exhaust ventilation over workstations
- Small quantities purchased in plastic containers for ease of handling and safety
- Traps placed in floor drains

- Spill-absorbent bags available for emergencies
- Exposures immediately reported
- Spill clean up accomplished by properly trained and equipped staff

PPE   Avoid skin and eye contact. Use goggles, face shields, aprons, NIOSH-certified respirators, and boots in situations where spills and splashes are likely. Use appropriate protective gloves whenever hand contact is possible. Do not use latex examination gloves, as they are too fragile.

Medical Monitoring

Record preemployment baseline data for the respiratory tract, liver, and skin condition of any worker who will be exposed to formaldehyde. Thereafter, conduct periodic monitoring to detect symptoms of pulmonary or skin sensitization or effects on the liver.

Glutaraldehyde

In the past few years, many questions have been raised regarding the safe use of glutaraldehyde in the workplace. Glutaraldehyde is frequently confused with formaldehyde.

Hazard Location

Gluteraldehyde is used as a high-level disinfectant of semicritical devices in physicians' offices throughout the world.

Characteristics (Physical Properties)

Glutaraldehyde is toxic. It is also corrosive to eyes and skin and may cause permanent eye injury.

Potential Health Effects

Both eye and skin irritation can occur after direct contact with glutaraldehyde solution. The eye is particularly sensitive and is the single most important area to protect from splashes. Glutaraldehyde has also been associated with allergic contact dermatitis and asthma and is now considered to be a respiratory sensitizing agent.

Standards and Recommendations

OSHA does not have an enforceable PEL for this substance, nor has NIOSH developed its own REL for glutaraldehyde. However, in 1988, NIOSH adopted exposure criteria, which were established by the ACGIH, that state that a TLV of 0.05 ppm (0.2 mg/m^3) as an airborne exposure level should not be exceeded at any time during any part of the work shift. In the United Kingdom, the health and safety executive has recently recommended that glutaraldehyde exposures be kept below 0.05 ppm.

Exposure Control Methods

Some of the recommended precautions when working with gluteraldehyde are as follows:

- Install proper ventilation equipment. Each facility's requirements differ, depending on the amount of glutaraldehyde used, room size, and available ventilation.
- Perform vapor monitoring after installing ventilation equipment.

- Never wear latex gloves or surgical respirators; neither will provide adequate protection against glutaraldehyde.
- Wear only a NIOSH-certified respirator, such as a half-mask respirator using an organic vapor cartridge or a full-face cartridge respirator, to prevent inhalation problems. A full-face respirator will also offer eye protection.
- Wear appropriate type and length gloves, such as nitrile rubber, some surgical synthetics, and polyethylene. Obtain specific recommendation from the glove manufacturer before purchase.
- Wear protective eyewear, such as laboratory monogoggles (ie, tight-fitting safety goggles that are not penetrated by liquid splashes). Also consider wearing a face shield in conjunction with primary eye protection.
- Wear gowns or aprons that are capable of providing adequate protection from splashes and spills to prevent contact between glutaraldehyde and clothing.
- Make certain all containers and soaking bowls are sealed.
- Store containers away from heat and light.
- Avoid pouring or emptying into open areas such as sinks.
- Rinse carefully, with attention to transfer techniques that eliminate splashing.
- Follow the facility-written spill containment plan in the event of a spill.

Mercury

Mercury, also called "quicksilver," is another substance often found in medical practices. It is an odorless, silver-white metal that, if heated or exposed to room temperature, becomes an odorless gas. Elemental mercury is a metallic element that is liquid at room temperature.

Hazard Location
Some of the different types of equipment that contain mercury include:

- Batteries, including those for defibrillators, hearing aids, and pacemakers
- Thermometers
- Sphygmomanometers (blood pressure)
- Electrical instruments
- Lamps, including fluorescent, metal halide, high-pressure sodium, ultraviolet, and cathode-ray tubes
- Computers

Characteristics (Physical Properties)
Liquid mercury is a proven neurotoxin. Neurotoxins are chemicals that produce their primary toxic effects on the nervous system. While mercury salts are toxic by ingestion, the prime toxicity of liquid mercury is in the form of vapor.

Potential Health Effects
Mercury is toxic even in minute doses. Although inhalation is the major route of entry for mercury, the element can also be absorbed through the

skin. Exposure to short-term high levels of mercury can produce severe respiratory irritation, digestive disturbances, and marked renal damage. Exposure to mercury can cause damage even before symptoms occur.

Long-term exposure to low levels of mercury results in the classic mad hatter syndrome (ie, named for the makers of felt hats who used mercury in processing). Emotional instability, irritability, tremors, inflammation of the gums (ie, gingivitis), excessive salivation, anorexia, and weight loss characterize this syndrome. Mercury has also been reported as a cause of sensitization dermatitis.

Standards and Recommendations
The current OSHA PEL for mercury is 0.1 mg/m^3 as a ceiling value (Standard Number 29 CFR 1910.1000, Table Z-2). The NIOSH REL is 0.05 mg/m^3 averaged over an 8-hour work shift.

Pollution Prevention
Design a pollution prevention program using guidance from state and federal agencies. Elements of the program should include using protective clothing, training employees to carry out safe handling procedures, cleaning and calibrating mercury-containing equipment, and requesting a mercury vacuum sweeper and spill cleanup kit from a safety supply vendor.

If mercury spills are not promptly cleaned up, mercury may accumulate in the carpeting, on floors, and on other surfaces, such as porous laboratory sinks and counters. Mercury vaporizes easily at room temperatures.

Exposure Control Methods
There are many alternatives to the uses for mercury in equipment or as chemical compounds that are considerably less harmful. One example is using calibrated digital or gauge blood pressure cuffs instead of providing the pressure column in a sphygmomanometer. When implementing this change, the medical supply company providing the new cuffs may be able to dispose of the mercury-containing apparatus.

Engineering controls Emergency procedures for handling mercury contamination should include procedures for clean up as well as for respirator selection. Exhaust systems should be designed and maintained to prevent the accumulation or recirculation of mercury vapor into the workroom.

PPE Use disposable PPE such as shoe covers, protective gloves, special mercury vapor respirators, and gowns and hoods while cleaning up mercury spills.

Work practices Clean up spills promptly with special mercury vacuum cleaners, disposable PPE, and a water-soluble mercury decontaminant. Dispose of mercury waste according to EPA regulations (Standard Number 40 CFR 261.24).

Clearly post all spill areas until adequate clean up has been accomplished. If the spill is extensive, remove patients and employees other than the clean-up crew from the area.

Medical Monitoring
Record preexposure data for the respiratory tract, nervous system, kidneys, and skin of any worker who may be exposed to mercury. Periodically monitor urine mercury levels in workers who are routinely or accidentally exposed to this element. Although there is no critical level of mercury in urine that indicates mercury poisoning, observers have suggested that 0.1 mg to 0.5 mg of mercury/liter of urine has clinical significance.

Phenolics

Phenol is another chemical commonly used by health care workers in laboratory and treatment therapies. Phenolics were among the first disinfectants used in health care. Certain detergent disinfectants belong to the phenol group, including phenol, para-tertiary butylphenol (ptBP), and para-tertiary amylphenol (ptAP). They are generally used for a wide range of bacteria, but they are not effective against spores.

Hazard Location
Phenolics are widely used on floors, walls, furnishings, glassware, and instruments.

Characteristics (Physical Properties)
Phenol is considered to be very toxic to humans through oral exposure. Phenol is very soluble in water and is quite flammable.

Potential Health Effects
Some people may be able to detect the odor of phenol at a concentration of about 0.05 ppm. Serious health effects may follow exposure to phenol through skin absorption, inhalation, or ingestion. These effects may include local tissue irritation and necrosis, severe burns of the eyes and skin, irregular pulse, stertorous breathing (ie, harsh snoring or gasping sound), darkened urine, convulsions, coma, collapse, and death. Both ptBP and ptAP have caused health care workers to experience loss of skin pigment that was not reversed 1 year after use of the compounds was discontinued.

Standards and Recommendations
OSHA's PEL for phenol is 5 ppm (ie, 19 mg/m^3) as an 8-hour TWA (skin) (Standard Number 29 CFR 1910.1000, Table Z-1). NIOSH's REL for phenol is 20 mg/m^3 (ie, 5.2 ppm) for up to a 10-hour TWA with a 15-minute C of 60 mg/m^3 (ie, 15.6 ppm). Neither OSHA nor NIOSH has established exposure limits for ptBP or ptAP.

Exposure Control Methods
There are a number of exposure control methods to use when working with phenolics, and some of the methods are as follows:

Engineering controls Measures to control phenol exposure include process enclosure and local exhaust ventilation.

PPE When working with phenol, provide workers with and require them to use protective clothing (eg, gloves, face shields, splash-proof safety goggles) necessary to prevent any possibility of skin or eye contact with solid or liquid phenol or liquids containing phenol.

Work practices If there is any possibility that the clothing has been contaminated with phenol, an employee should change into uncontaminated clothing before leaving the work area and the suspect clothing should be stored in a closed container until it can be discarded or until provision is made for removal of the phenol. The person responsible for laundering or cleaning the clothes should be informed of phenol's hazardous properties.

Immediately wash skin that becomes contaminated with phenol with soap or mild detergent and rinse with water. Do not allow eating and smoking in areas where solid or liquid phenol or liquids containing phenol are handled, processed, or stored. Employees who handle solid or liquid phenol or liquids containing phenol should wash their hands thoroughly with soap or mild detergent and water before eating, smoking, or using toilet facilities.

WORKING WITH CHEMICALS

Chemicals can affect the body through contact, absorption (ie, spills on unbroken skin, touching, absorption through the pores), inhalation, injection (ie, needle sticks, punctures, being cut by a sharp object), or ingestion (ie, swallowing). What is often overlooked is that many chemicals emit gases or fumes, and these can settle onto clothing, beverages, food, hair, or follicles or anything else in the area. Even if there is no immediate contact, exposure can still occur. Table 4.1 presents ideas for safe work habits.

Mild or short-term exposure can cause headaches, watery eyes, a sore throat, dizziness, skin rashes, an upset stomach, tiredness, or shortness of breath. Longer exposures can cause cancer; lung, kidney, liver, and heart disease; birth defects; miscarriages; sterility or infertility; and brain or nervous system damage.

Possible interactions may occur as a result of the multiple exposures that exist in a medical practice. These interactions may involve (1) exposures to chemical and/or physical agents; (2) an individual's use of tobacco, alcohol, or drugs; or (3) the physiological or psychological state of the worker. To determine an exposure, it is imperative to consider these other possible exposures or factors that might influence the results.

New and less hazardous chemicals are regularly introduced into the marketplace. Use substitutions for harmful chemicals whenever possible. For example, there are alternatives to glutaraldehyde. These include new high-level disinfection processes as well as less-hazardous chemicals. However, even with substitutions, it is still necessary to apply appropriate worker protection and safe work practices.

Some chemicals have characteristics that can be detected by employees and serve as a warning of the chemicals' presence. The most common warning is odor. The lowest concentration at which the odor of a chemical can be detected is called the odor threshold. For example, formaldehyde has a very distinctive odor at a very low airborne concentration. Because not all substances have an odor (eg, mercury) there would be no warning odor to indicate their presence.

TABLE 4.1

Practice Safe Work Habits

- Take only as much as you need from the container, then close it.
- Follow label precautions regarding dilution. Don't make the concentration too strong.
- Make certain the area is well ventilated and try to avoid inhaling vapors.
- Properly label all portable containers.
- Use gloves. Inspect all PPE before you use it. Afterward, remove PPE carefully so that you don't contaminate yourself.
- Wash thoroughly with soap and water after the job is done and before eating or using the toilet.
- Dispose of extra product and wastes carefully.

Reproduced with permission from 2000 Business & Legal Reports, Inc, OSHA Required Training for Supervisors.

Repeated exposure to a chemical can cause olfactory fatigue, preventing a worker from continuing to smell the chemical. Also, not all people detect odors equally well. For instance, long-term smokers of tobacco frequently notice a diminution of smell. For this reason, utilizing only the sense of smell is unreliable.

Another method of detecting the exposure level of a chemical is by biologically monitoring blood, urine, or exhaled air. This can determine just how much of a chemical has entered an employee's body.

It may be necessary to consult with an industrial hygienist. NIOSH may also provide consultation through its Health Hazard Evaluation Program. These specialists may be able to determine the source of health complaints and measure the amount of toxic chemicals in the air or office environment. Collecting air from the employee's breathing zone is an example of the type of testing performed. Those concentrations are then compared to guidelines for that chemical.

Multiple Chemical Sensitivity (MCS)

Employees react differently to chemical exposures. There are a range of symptoms associated with exposure and sensitization to chemicals used in medical practices. For example, some chemicals are respiratory irritants and can lead to the development of occupational asthma and sensitization in certain individuals. Other chemicals may blister the skin while still others may cause reproductive problems for females and unborn fetuses. Some employees are far more sensitive to a chemical's properties than others (eg, its odor or the fact that it is an eye irritant). Training must encourage employees to report adverse reactions to chemicals.

For some people the sensitization is not a serious problem. They may have what appears to be a minor allergy to one or more chemicals. Other people, however, are much more seriously affected. They may feel tired all the time and suffer from mental confusion, breathing problems, sore muscles, and a weakened immune system. Such people refer to this problem as multiple chemical sensitivity (MCS).

MCS is a disorder that appears to be triggered by exposures to chemicals in the environment. Individuals with MCS can have symptoms from chemical exposures at concentrations far below the levels tolerated by most people. Symptoms occur in more than one organ system in the body, such as the nervous system and the lungs. Exposure may be from the air, from food or water, or through skin contact. The symptoms may look like an allergy because they tend to come and go with exposures, though some individual's reactions may be delayed. A number of investigators believe that MCS represents a generalized immune system response. As MCS gets worse, reactions become more severe and increasingly chronic, often affecting more bodily functions. No single widely available medical test can explain symptoms.

In the early stages of MCS, repeat exposure to the substance or substances that caused the initial health effects provokes a reaction. After a time, it takes less and less exposure to this or related chemicals to cause symptoms. As the body breaks down, an ever-increasing number of chemicals, including some unrelated to the initial exposure, trigger a reaction.

Good indoor air quality and substitution of less toxic materials in the workplace help prevent chemical exposures that affect the health of chemically sensitive employees. Employers should take complaints about indoor air problems seriously.

If all personal and environmental controls are in place and skin and/or respiratory irritation persists, employees should consult their supervisor about possible reactions to various chemicals in the workplace. For example, some health care workers may have an allergic reaction to latex gloves. Hypersensitivity reactions include contact urticaria, rhinitis, conjunctivitis, anaphylactic reactions, and asthma. Some of these are attributable to local contact with glove latex while others are from exposure to latex allergens, transferred from gloves to powder, that may become airborne during handling. The hypersensitivity reactions to airborne latex allergens may be experienced not only by those wearing gloves, but also by sensitive workers in the immediate vicinity. Some of these effects caused by latex could be attributed mistakenly to another chemical. Therefore, those complaining of hypersensitivity reactions in clinics using the chemical should fully investigate all possible chemical sources of their complaints. It is important to accurately identify the exact source of ill health to be certain that the proper corrective measures are taken.

Hazardous Chemicals in Laboratories

Most medical practices do not have laboratories, and when they do, bulk chemicals are not usually used. Testing kits are usually used, which are exempt from both the Hazard Communication Standard (HAZCOM) (Standard Number 29 CFR 1910.1200) and Laboratory Safety Standard (Standard Number 29 CFR 1910.1450) requirements.

In laboratories covered by the Clinical Laboratory Improvement Amendments of 1988 (CLIA) or limited-scope CLIA laboratories that use bulk hazardous chemicals for analysis, employees may be exposed to hazardous chemicals, such as formaldehyde, mercury, and phenol. Accidents, injuries, and exposures may occur in these laboratories, resulting in chemical-related illnesses ranging from skin and eye irritation to fatal pulmonary edema.

In CLIA or limited-scope CLIA laboratories, the use of chemicals is generally limited to small quantities. Therefore, laboratory workers may be exposed to many different chemicals, but generally in small quantities for short periods of time.

On January 31, 1990, OSHA issued the Laboratory Safety Standard to address the hazardous substances generally found in large laboratories. That standard establishes the requirement for developing and implementing a Chemical Hygiene Plan covering laboratories using hazardous chemicals. It also contains all of the elements of HAZCOM, as well as a number of elements specific to the laboratory environment. Because the Laboratory Safety Standard is a specialized standard specific to laboratories, this book does not cover the standard. The laboratory rule requires continued compliance with OSHA's PELs and with the employer's written Chemical Hygiene Plan.

The written Chemical Hygiene Plan is the core of the standard and affords flexibility in providing the type of employee protection appropriate for a specific workplace. This plan, which the employer develops, specifies the training and information requirements of the standard. It also establishes appropriate work practices, standard operating procedures, methods of control, measures for appropriate maintenance and use of PPE, medical examinations, and special precautions for work with particularly hazardous substances. Some medical practices may be able to meet certain requirements of the written plan by relying on their existing safety and health plans. The employer must evaluate the effectiveness of the plan at least annually and update it as necessary. The written program must be available to employees.

OSHA HAZARD COMMUNICATION STANDARD

Because of the seriousness of safety and health problems that can occur in the workplace and because many employers and employees know little or nothing about these hazards, OSHA issued a rule in 1983 called Hazard Communication, which applied to employers in the manufacturing sector of industry. The rule is commonly known as the Hazard Communication Standard (HAZCOM) or the "Right-to-Know Law." The scope of the rule was expanded in 1987 to include employers in the non-manufacturing sector, including medical practices, nursing homes, hospitals, and other health care facilities.

The standard's basic goal is to ensure that employers and employees know about chemical hazards and how to protect themselves. This knowledge, in turn, should help to reduce the incidence of chemical source illnesses and injuries.

What the Standard Requires

HAZCOM establishes uniform requirements to assure that the hazards of all chemicals imported into, produced by, or used in US workplaces are evaluated and that the resultant hazard information and associated protective measures be transmitted to affected employers and potentially exposed employees via container labels and material safety data sheets (MSDSs). The MSDS is a detailed information bulletin prepared by the manufacturer or importer of a chemical that describes the chemical. Table 4.2 provides an MSDS checklist. In addition to providing the detailed chemical information, all covered employers must have a hazard

TABLE 4.2

Material Safety Data Sheet Checklist

Information to be included in each MSDS:

_____ 1. Product or chemical identity used on the label.
_____ 2. Manufacturer's name and address.
_____ 3. Chemical and common names of each hazardous ingredient.
_____ 4. Name, address, and telephone number for hazard and emergency information.
_____ 5. Preparation or revision date.
_____ 6. The hazardous chemical's physical and chemical characteristics, such as vapor pressure and flashpoint.
_____ 7. Physical hazards, including the potential for fire, explosion, and reactivity.
_____ 8. Known health hazards.
_____ 9. OSHA PEL, ACGIH TLV, or other exposure limits.
_____ 10. Emergency and first-aid procedures.
_____ 11. Whether OSHA, NTP, or IARC lists the ingredient as a carcinogen.
_____ 12. Precautions for safe handling and use.
_____ 13. Control measures such as engineering controls, work practices, hygienic practices, or PPE required.
_____ 14. Primary routes of entry.
_____ 15. Procedures for spills, leaks, and clean-up.

communication program to provide this information to their employees through labels on containers, MSDSs, and training.

HAZCOM ensures that all employers receive the information they need to properly inform and train their employees and to design and enact employee protection programs. It also provides necessary hazard information to employees so they can participate in and support the protective measures in effect at their workplaces.

When using hazardous chemicals, it is important to employ appropriate industrial hygiene practices, such as wearing the proper PPE and installing environmental controls. Every health care worker who uses hazardous chemicals should be well trained and knowledgeable about the properties of the material(s), their potential harmful effects, and the work procedures of the facility. Appropriate gloves, gowns, and eye protection should be available and worn. If necessary, additional ventilation should be installed to keep vapor concentrations below the PEL, and respiratory protective equipment should be available for emergency situations. All employees should be aware of spill clean-up procedures, which should include written instructions.

Inspection Procedures

OSHA randomly conducts inspections to ascertain the degree of compliance with HAZCOM. According to OSHA, failure to comply with HAZCOM is among the most frequent reasons for issuing a citation.

It has been repeatedly demonstrated that OSHA takes the issues of chemicals in the workplace and HAZCOM very seriously. Inspections may be announced or unannounced, quite thorough, and conducted by Compliance Safety and Health Officers (CSHOs). Usually, inspections occur for five primary reasons:

1. Employee complaints
2. Fatalities
3. Routine inspections for high-hazard facilities (eg, chemical, manufacturing, construction companies)
4. Multiple (ie, three or more hospitalized) injuries from the same accident
5. General inspections

State "Right-to-Know Laws"

Individual states may implement their own "Right-to-Know Laws," but they must meet or exceed the federal HAZCOM regulations.

In many states, it is not the federal OSHA that oversees compliance, but the individual states' OSHA rules. Under these circumstances, the laws must be at *least* as stringent as the federal regulations. In many cases, they are even more stringent.

How to Comply with the Standard

The following six steps will aid in complying with the standard and developing the practice's hazard communication program:

1. *Read the standard.*
   - Make sure the provisions of the standard are clear and understood.
   - Know the employer's responsibility.

2. *List the hazardous chemicals in the workplace.*
   - Walk around the workplace, read all container labels, and list the identity of all materials that may be hazardous; the manufacturer's product name, location, and telephone number; and the work area where the product is used.
   - Check to ensure that all hazardous chemicals purchased are included on the list.
   - Review the list and determine whether any substances are exempt.
   - Establish a file on hazardous chemicals used in the workplace, and include a copy of the latest MSDSs and any other pertinent information.
   - Develop procedures to keep the list current. When new substances are used, add them to the list.

3. *Obtain MSDSs for all chemical substances.*
   - If there is not an MSDS for a hazardous substance in the workplace, request a copy from the chemical manufacturer, distributor, or importer as soon as possible. MSDSs can be obtained via the Internet or by writing to the chemical manufacturer. An MSDS must accompany or precede the shipment and must be used to obtain identifying information such as the chemical name and the hazards of a particular substance.

- Review each MSDS to be sure that it is complete and clearly written. The MSDS must contain the physical and chemical properties of a substance, as well as the physical and health hazards, routes of exposure, precautions for safe handling and use, emergency and first-aid procedures, and control measures.
- If the MSDS is incomplete or unclear, contact the manufacturer or importer to get clarification on the missing information.
- Make sure the MSDS is available to employees.

Because some of the terminology on an MSDS may not be part of the employees' everyday conversation, it can be intimidating. To some, it can seem entirely too much to try to assimilate. It is essential that all employees know what the terminology means and how to interpret it. Their health and safety depend upon this information. Implement refresher quizzes, flash cards, or any other means of training, if necessary.

How MSDSs are stored varies from practice to practice. One simple way is to place the MSDSs into a three-ring binder. Smaller practices may prefer to keep the sheets in a file drawer in a folder with the written program.

Because all MSDS forms have not been standardized, some medical practices transfer the information onto a uniform format for ease of use. Once the task is accomplished, it is doubtlessly simpler for quick reference, but transferring the information can be quite time-consuming and must be precisely accurate. Due to the assumption of liability for the information that is transferred from the original MSDS to a medical office form, this practice is not recommended.

4. *Make sure that all containers are labeled.*

The manufacturer, importer, or distributor is responsible for labeling containers, but the employer must adhere to the following:

- Ensure that all containers of hazardous chemicals in the workplace are labeled, tagged, or marked and include the identity of the hazardous chemical and the appropriate hazard warnings. Container labels for purchased chemicals must also include the name and address of the chemical manufacturer, importer, or other responsible party.
- Check all incoming shipments of hazardous chemicals to be sure that they are labeled.
- If a container is not labeled, obtain a label or the label information from the manufacturer, importer, or other responsible party or prepare a label using information obtained from these sources. Employers are responsible for ensuring that containers in the workplace are labeled, tagged, or marked.
- If the existing label is removed or defaced, be sure to immediately mark the container with the required information.
- Instruct employees on the importance of labeling secondary containers into which they have poured hazardous chemicals.

5. *Develop and implement a written hazard communication program.*
   The program must include the following:
   - Container labeling and other forms of warnings
   - MSDSs
   - Employee training based on the list of chemicals, MSDSs, and labeling information
   - Methods for communicating hazards and protective measures to employees and others

6. *Train employees.*
   Employers must establish a training and information program for employees who are exposed to hazardous chemicals in their work area at the time of initial assignment and whenever a new hazard is introduced into their work area. Document the HAZCOM training that employees receive, including names and dates of training.

   Discussion topics during training must include the following:
   - HAZCOM and its requirements
   - The components of the hazard communication program in the employees' workplaces
   - Operations in work areas where hazardous chemicals are present
   - Where the employer keeps the written hazard evaluation procedures, communications program, lists of hazardous chemicals, and the required MSDS forms

   Employee training must consist of the following elements:
   - How the hazard communication program is implemented in the workplace, how to read and interpret information on labels and the MSDS, and how employees can obtain and use the available hazard information.
   - The hazards of the chemicals in the work area. (The hazards may be discussed by individual chemical or by hazard categories such as flammability.)
   - Measures that employees can take to protect themselves from the hazards.
   - Specific procedures put into effect by the employer to provide protection, such as engineering controls, work practices, and the use of PPE.
   - Methods and observations, such as visual appearance or smell, that employees can use to detect the presence of a hazardous chemical to which they may be exposed.

All employees have a right to a safe work environment, and the controls provided in this chapter help employers gain the knowledge they need to attain that goal. It is the responsibility of employees to perform tasks safely. Employees must realize that if they fail to work safely, they are endangering not just their own health, but the health and safety of their coworkers as well.

ENDNOTES

1. Jordan SLP. The correct use of glutaraldehyde in the healthcare environment. *Gastroenterology Nursing*. 1995;18(4).

2. National Institute for Occupational Safety and Health. *Hazard evaluation and technical assistance report: St. Francis Hospital, Honolulu, HI*. Cincinnati, OH: US Department of Health and Human Services, Public Health Service, Centers for Disease Control and Prevention, National Institute for Occupational Safety and Health, NIOSH Report No. HETA 82-303-1271, NTIS No. PB-84-209-444/A02.

3. National Institute for Occupational Safety and Health. Current Intelligence Bulletin 34: *Formaldehyde; Evidence of Carcinogenicity*. Cincinnati, OH: US Department of Health and Human Services, Public Health Service, Centers for Disease Control and Prevention, National Institute for Occupational Safety and Health, DHHS (NIOSH) Publication No. 81-111, NTIS No. PB-83-101-766.

chapter 5

Other Health and Safety Issues

If you start to take Vienna, take Vienna.
Napoleon Bonaparte (1769–1821)

Other health and safety issues for a medical practice include pharmaceuticals and controlled substances, medical gas handling, safe maintenance of equipment, and injuries from slips, trips, or falls. This chapter looks at all these areas of concern.

PHARMACEUTICALS AND CONTROLLED SUBSTANCES

Advances in health care have occurred in large part due to the development of a vast array of pharmaceuticals and their many exciting medical discoveries. Health care professions assisted by drugs are successfully fighting some of the world's most fearful and persistent diseases. Pharmaceutical research activities now conduct clinical trials with active patient participation and education.

New drug applications (NDAs) have dramatically increased for products that are expected to play key roles in the improved understanding and management of the various diseases and illnesses. These advances in pharmaceutical development, coupled with an explosion of new information about illnesses and diseases and in some cases their genetic origin, has brought about advances in the treatment of cancer, heart disease, mental illness, and pain management.

Unfortunately, as is the case with any new technology, these advances present a mixed blessing. Some pharmaceutical drugs can be considered "a wolf in sheep's clothing." Health care professionals, in their day-to-day activities, see the healing side of pharmaceuticals. However, law enforcement officers become familiar with the addictive side and self-destruction caused by some pharmaceuticals. To police, these pharmaceuticals are drugs, and to health care professionals, pharmaceuticals are medication. When most people think of medicine, they think of how it helps people and not the danger and destruction that it can inflict. Pharmaceutical drugs are intended to be used for medical and therapeutic purposes by a practitioner, and, out of several thousand drugs, only a handful are abused because of their narcotic or mood-altering effect. Drugs of abuse include narcotics, stimulants, amphetamines, depressants, barbiturates, and benzodiazepines. For example, Ritalin, which is used for hyperactive children, has the opposite effect when used by adults. It is one of the stronger "speed drugs" available on the market. Ritalin is extremely popular on college campuses, and in fact is referred to by students as "Vitamin R."

Pharmaceutical Diversion

While most medical practices that handle pharmaceutical drugs are not typically the target of forcible theft, they sometimes become a target because they store, prescribe, or dispense narcotic and mood-altering drugs.

Pharmaceutical drug abusers are usually con artists and may approach a medical practice on the pretense of being a legitimate patient. This type of patient may even resort to stealing prescription blanks off the desks of receptionists and doctors and forging their own prescriptions or increasing the amount of a certain drug on an original prescription. Pharmaceutical drug abuse is increasing across the nation, as it is significantly safer and cheaper for the abuser to obtain a legal prescription than to obtain the drugs on the streets and risk being shot or robbed. Furthermore, the probability of being caught by a physician or pharmacist and arrested is less. Additional reasons for pharmaceutical drug abuse include:

- The desire to decrease the risk of infection due to the use of shared needles by abusers.
- The quality of drugs is consistent. Pharmaceuticals are the same strength nationwide.
- The environment from which to obtain the drugs is relatively violence-free.
- Insurance often pays for the prescription drugs.
- Drug abusers are becoming more sophisticated.
- There are large profits derived from selling pharmaceutical drugs on the streets.

Drug abusers employed in a medical practice have easier access to drugs, either by convincing a physician to write a prescription or by theft of drugs or prescription pads. Employees who abuse pharmaceutical drugs typically falsify medical records in order to get the drugs they want. For example, medications intended for a patient to alleviate pain or to use as a sedative prior to surgery may be diverted for the care provider's personal use. Or in some cases, the medication is replaced with another substance or another drug presenting the patient with a potentially dangerous risk.

Physicians and staff who are drug abusers present additional concerns. They may be providing or assisting with medical treatment, such as minor surgery or physical exams, and, if under the influence of drugs, they may be unable to properly conduct themselves. They become a danger to the patient and other staff because they may be drowsy, dizzy or lightheaded, nauseous, weak, or lose consciousness. Some drugs of abuse cause tremors or hallucinations that impair a care provider's ability to properly care for the patient. Even the patient information provided with prescriptions includes precautions that advise not to operate machinery or drive when taking certain medications. Figure 5.1 shows sample warning messages found on a pharmaceutical drug and its physician-recommended dosage.

FIGURE 5.1

Sample Patient Prescription Information

Reproduced with permission from *Patient Prescription Information,* published by CVS Pharmacy.

> DO NOT DRINK ALCOHOLIC BEVERAGES WHEN TAKING THIS MEDICATION.
>
> MAY CAUSE DROWSINESS.
>
> ALCOHOL INTENSIFIES EFFECT.
>
> USE CARE USING MACHINES.
>
> DO NOT "DOUBLE-UP" THE DOSES.
>
> DO NOT INCREASE THE DOSE OR TAKE THIS MORE FREQUENTLY THAN PRESCRIBED.
>
> BECAUSE THIS MEDICATION MAY CAUSE DROWSINESS, USE CAUTION PERFORMING TASKS REQUIRING ALERTNESS.
>
> THIS IS A STRONG COUGH MEDICATION AND MUST ONLY BE USED FOR THE CONDITION AND BY THE PERSON FOR WHICH IT IS PRESCRIBED. IT SHOULD NOT BE USED FOR JUST ANY COUGH.

Another well-recognized problem is the development of addiction to prescription drugs that are prescribed for legitimate purposes. In a 1995 Drug Abuse Warning Network (DAWN) study involving 489 emergency rooms, patients cited dependence as the number one reason for drug abuse.

Solutions

Awareness of the scope and seriousness of the problem is the first step toward a solution. Medical practices and clinics may have unique advantages available in dealing with the problem of pharmaceutical diversion, such as committing professional management and staff to deal with the problem, providing established accounting procedures, and training and education opportunities through pharmaceutical suppliers, medical associations, licensing authorities, and police agencies. When a medical practice stocks pharmaceutical drugs or provides prescriptions, it must have strict controls in place to reduce drug diversion. Such drug diversions are as follows:

1. Employment policies should include procedures addressing pharmaceutical drug diversion. What the policies include and how they are implemented is key to their success. Policies must be uniform for all employees, ongoing, and be customized to fit the particular medical practice operation. Two options for preventing drug diversion among employees include:

   - Preemployment background investigations. These investigations should include the work history of the applicant, verifying licensing and qualifications, and a check for prior drug offenses.

- Drug testing. If a medical practice initiates drug testing, several considerations are important. The policy must be well thought out and functional for the particular medical practice. For example, drug testing must include emergencies, random testing, and ongoing testing.

2. Training, which encompasses information about the nature and extent of the problem, how to recognize attempts to obtain drugs illegally, how to recognize the signs and symptoms of drug abuse, and the security measures that must be taken to prevent the diversion of the practice's pharmaceutical supplies, is an important element for reducing pharmaceutical drug diversion.
3. Although regulations for controlling pharmaceutical drugs help reduce the potential for diversion and abuse, security of drugs within the medical practice is the most effective control. Federal drug laws and regulations cover controlled substances. There must be controls in place with limited access for the staff and no access for unauthorized individuals. The following precautions should be taken to minimize the risk of theft or unauthorized use of pharmaceutical drugs:

- Store all pharmaceutical drugs, including samples, in a locked area with controlled access.
- Initiate and monitor a program of control, which should include stocking, tracking, and disposing of the drugs. The program should also include documentation of the disposal method and must comply with applicable federal and state laws.
- Take part in services provided by pharmaceutical companies such as computerized stock control, tracking systems, and management of returns.
- Make sure that all pharmaceuticals are stored in conformance with federal and state laws.
- Do not label the controlled substances cabinet.
- Keep syringes and needles out of sight.
- Store prescription pads and order forms, including DEA Form 222, in a secured area with controlled access. Limit access to these forms and drugs to specific management-approved staff.

Pharmaceutical drug safety includes a two-pronged approach for health care professionals—due diligence and education, with a major thrust toward education.

MEDICAL GAS HANDLING AND STORAGE

Medical gases (ie, compressed gases) are supplied in cylinders. The most common compressed gases used in medical offices include nitrous oxide, carbon dioxide, medical air, oxygen, nitrogen and nitrogen NF (ie, medical-grade nitrogen), liquid nitrogen, and some specialty gases (eg, argon) for lasers. They can be flammable, aid combustion, toxic, explosive, poisonous, corrosive, or all of these. For example, oxygen and nitrous oxide are labeled as nonflammable, but they are oxidizing agents and can support combustion and fire. Nitrogen, carbon dioxide, and argon are toxic to use because they displace the air we breathe and can cause asphyxiation. Therefore, anyone who handles or uses compressed gases must:

1. Be properly trained.
2. Receive and understand all relevant safety literature, including Material Safety Data Sheets (MSDSs).
3. Follow the manufacturer's or supplier's operating instructions.
4. Read, understand, and apply safety precautions on gas cylinder labels.
5. Observe all applicable safety codes when installing gas equipment.

Cylinder Safety

Because compressed gases are under extreme pressure, the cylinders must always be handled with care. An exploding cylinder can have the same destructive effect as a bomb. Proper handling requires training and a well-enforced safety policy. Some safety standards and recommendations for the safe handling of compressed gas cylinders include:

- NIOSH recommendations for storing and handling compressed gas cylinders.
- OSHA Regulation, Subpart H—Hazardous Materials (Standard Number 29 CFR 1910.101-106).
- ANSI-Z48.1, *Marking Portable Compressed Gas Containers.*
- Compressed Gas Association (CGA) Pamphlets (Incorporated by reference into OSHA Regulation [Standard Number 29 CFR 1910, Subpart H]).
- NFPA 53, *Recommended Practice on Materials, Equipment, and Systems Used in Oxygen-Enriched Atmospheres.*
- Provide and use MSDSs.
- Some state "Right-to-Know Laws" cover compressed gases.
- Medical gas vendors usually provide free safety literature upon request.

Handling and Storing

Safe and proper handling and storing can prevent injuries. The following are guidelines for general cylinder safety, whether handling or storing compressed gas:

- Never place a cylinder of one type against a cylinder of another type.
- Do not allow cylinders to be dropped or to strike each other.
- Never subject compressed gas cylinders (full or empty) to temperatures in excess of 125°F.
- Do not allow flame to come into contact with any part of a compressed gas cylinder.
- Store, transport, and use cylinders with the labels to the outside for easy product identification.
- Store and transport cylinders with safety caps in place (if appropriate).
- Ensure that each new tank is equipped with the proper washer between the stem and the yolk.

- Do not expose cylinders to dampness, corrosives, or abrasive surfaces.
- Store or handle cylinders with protective caps screwed on, down to the last thread.

The following section provides recommendations for cylinder safety. Table 5.1 shows gases and cryogenic liquids organized in product categories having similar health hazards and storage requirements.

PPE

A cylinder striking someone or pinching a finger, toe, or other extremity is a common cause of injury. The type of PPE required depends on the products being handled at the particular medical practice. Minimum PPE required when handling cylinders and containers includes:

- Gloves to protect the hands against common pinching injuries.
- Safety glasses to protect the eyes against injuries associated with pressure release.
- Safety shoes with metatarsal supports to protect against foot injuries from falling cylinders.

Receiving Cylinders and Containers

When receiving cylinders and containers, the medical practice should:

- Read the cylinder labels to be sure that the gas is what was ordered. Never rely on cylinder color for product identification.
- Thoroughly inspect the cylinders for any obvious damage. The cylinder surface should be clean and free from defects, such as cuts, gouges, burns, and obvious dents. Such damage could weaken the cylinder, creating a danger of failure, or it could make the cylinder unstable and more likely to tip over. Make sure the cylinder stands steady on its base and does not wobble.
- Cylinders with neck threads should have a cap in place over the valve. Remove the cap by hand or with a wrench designed solely for the purpose of removing cylinder caps. Never use a screwdriver, crowbar, or other leverage device to remove the cap. Doing so could accidentally open the valve or damage it.
- Check the cylinder valve to be sure it is not bent or damaged. A damaged valve could leak, fail, or not connect tightly enough. Make sure the valve is free of dirt and oil, which could contaminate the gas. Dirt particles propelled in a high-velocity gas stream could cause a spark, igniting a flammable gas. Oil and grease can react with oxygen and other oxidizers, causing an explosion.
- If any cylinder is received with missing or unreadable labels and markings, visible damage, an unstable base, a missing cap, or a bent, damaged, or dirty valve, do not use the cylinder. Contact the supplier and ask for instructions.

Testing for Leaks

After completing the inspection of all cylinders and containers as described above, a leak test should be conducted using an approved leak test method. If leakage is detected, follow the supplier's procedures for handling leaking cylinders. If no leakage is detected, secure the cylinder

TABLE 5.1

Sample of Gases and Liquids with Similar Health Hazards and Storage Requirements

Table 1: Flammable Gases

Store these products in well-ventilated areas equipped with electrical equipment in accordance with Article 500 of the National Electrical Code. These products may also be stored in covered outdoor areas, but keep them off the ground and away from radiant heat sources. Keep all ignition sources and combustible materials out of the storage area. Cylinder temperature should not exceed 125°F (52°C).	1,3-Butadiene 1-Butene Acetylene Butane cis-2-Butene Deuterium Difluoromethane Dimethyl ether Dimethylamine Ethane Ethyl chloride Ethylene	Hydrogen Isobutane Isobutylene Methane Methyl chloride Methyl fluoride Methylamine Natural gas Propane Propylene trans-2-Butene Trimethlyamine	**Note:** Flammable, noncorrosive, low-toxicity products can be safely stored together with compounds listed in Table 4, Section B: Toxic, Flammable.

Table 2: Pyrophoric Gases

Store these products in well-ventilated areas or in covered outdoor areas, but keep it off the ground and away from radiant heat sources. Keep combustible materials out of the storage area. Cylinder temperatures should not exceed 125°F (52°C).	Disilane Methylsilane Silane	**Note:** These products cannot be stored with any other products.

Table 3: Nonflammable Gases

Store these products in well-ventilated areas or in covered outdoors areas, keeping them off the ground and away from radiant heat sources. Cylinder temperatures should not exceed 125°F (52°C).	1, 1, 1, 2-Tetrafluoroethane Air Ammonia Argon Bromotrifluoromethane Carbon dioxide Chlorodifluoromethane Chloropentafluoroethane Helium Hexafluoroethane Krypton	Neon Nitrogen Nitrogen trifluoride Nitrous oxide Octafluorocyclobutane Octafluoropropane Oxygen Pentafluoroethane Sulfur hexafluoride Tetrafluoromethane Trifluoromethane Xenon	**Note:** Nonflammable, noncorrosive, low-toxicity products can be safely stored together with the compounds listed in Table 4, Section C: Toxic, Nonflammable.

TABLE 5.1
continued

Table 4: Sections A, B, and C

These products are very toxic. The slightest exposure to these products can kill. Preferably, store these toxic gases outdoors in a fenced-in area that has been locked and posted. This area must be covered, with the cylinders kept off the ground and away from radiant heat sources. If stored indoors, toxic gas cylinders must be kept in well-ventilated areas. In addition to being toxic, arsine, diborane, and phosphine are also flammable. The storage area for flammable toxic gases should have electrical equipment conforming to Article 500 of the National Electrical Code. Keep all ignition sources and combustible materials out of storage areas.	**Table 4, Section A: Highly Toxic Gases**		**Note:** These products cannot be stored with any other products. The gases listed in the three sections of Table 4 may be stored together.
	Arsine	Nitrogen dioxide	
	Diborane	Phosphine	
	Hydrogen selenide	Silicon tetrafluoride	
	Nitric oxide		
	Table 4, Section B: Toxic, Flammable Gases		
	Carbon monoxide	Germane	**Note:** These products can be stored with other flammable products (Table 1). The gases listed in the three sections of Table 4 may be stored together, if space is limited.
	Carbonyl sulfide	Hydrogen sulfide	
	Dichlorosilane	Methyl mercaptan	
	Ethylene oxide	Trimethylboron	
	Table 4, Section C: Toxic, Nonflammable Gases		
	Boron trichloride	Hydrogen chloride	**Note:** These products can be stored with other nonflammable products (Table 3). The gases listed in the three sections of Table 4 may be stored together, if space is limited.
	Boron trifluoride	Methyl bromide	
	Boron-11 trifluoride, enriched	Phosphorus pentafluoride	
	Chlorine	Sulfur dioxide	
	Chlorine trifluoride	Sulfur tetrafluoride	
	Germanium tetrafluoride	Tungsten hexafluoride	
	Hydrogen bromide		

Table 5: Liquid Products

Store these products in well-ventilated areas equipped with electrical equipment in accordance with Article 500 of the National Electrical Code. These products may also be stored in covered outdoor areas, but keep them off the ground and away from radiant heat sources. Keep all ignition sources and combustible materials out of the storage area. Cylinder temperature should not exceed 125°F (52°C).	1-Pentene	Octane	**Note:** Liquid products cannot be stored with gaseous products.
	Hexene	n-Pentane	
	Hydrogen fluoride	Silicon tetrachloride	
	Isooctane		
	Isopentane	Tetraethyl silicate	
	Methanol	Toluene	
	n-Heptane	Trichlorosilane	
	n-Hexanen-	Xylenes	

Reproduced with permission from *Storing Cylinders and Containers*, published by Praxair Technology, Inc. Copyright 1999, 2001 Praxair Technology, Inc. All rights reserved. For more information visit www.praxair.com.

valve cap in place before moving the cylinder to the point of use or to the storage area. It is normal for cryogenic liquid containers, such as liquid nitrogen, to vent through their relief valves to relieve excess pressure caused by temperature or pressure changes. This venting is normal and is not a leak.

Moving Cylinders and Containers

Cylinders and containers must always be carefully moved. Mishandling can result in a damaged valve or ruptured cylinder, which can expose personnel to the hazards associated with these gases. Before moving a cylinder to the storage area or point of use or before returning the cylinder to the supplier, ensure that:

- The outlet valve is fully closed.
- The outlet valve dust plug or pressure cap is tightly affixed for cylinders equipped with these protection devices.
- The valve protection cap is properly secured on cylinders with neck threads. Note: Valve caps must always be in place while moving or transporting cylinders or when they are in storage.

While moving full or empty cylinders, medical practices should:

- Always use carts or hand trucks designed for this purpose. The most common hand trucks used in medical practices are the four-wheel hand trucks used to transport a single cylinder up to 12 inches in diameter and the medical E-size cylinder cart for transport of medical oxygen cylinders.
- Never carry, drag, slide, or roll cylinders.
- Never drop cylinders or allow them to violently strike each other.
- Never lift cylinders by the cap or with a lifting magnet.

After moving a cylinder to its point of use, secure the cylinder in place. Use cylinder stands, clamps, or other securing devices that are recommended by the supplier.

Storing Cylinders and Containers

NFPA and OSHA codes govern storage of compressed gas cylinders and cryogenic liquid containers. Local codes may also apply. Medical practices should know and obey all codes governing storage.

In general, store cylinders so that they can't be easily toppled over. Remember, danger exists not only from accidental release of gas from cylinders that are damaged in a fall, but also from their striking someone and causing injury. To avoid accidents when storing cylinders and containers, remember:

- Store cylinders upright in compact groups, interlocking them so that each cylinder physically contacts those around it. Never stack them like cords of wood.
- Cylinders should be secured to the wall with a safety chain.
- Do not stand cylinders loosely or in a haphazard manner. A single cylinder that topples over can create a domino effect, causing other cylinders to fall.
- Single cylinders should be secured in place or on a cylinder cart so that they can't be readily knocked over.

- Keep stored cylinders out of high-traffic areas. Do not store them near the edges of platforms.
- Avoid storage in areas where there are activities that could damage or contaminate the cylinders.
- Electric arc welding can destroy the integrity of cylinder metal if a welder carelessly strikes the cylinder.
- Overhead hoists can drip oil or grease onto cylinders, thus contaminating them.
- Never store cylinders with flammable materials.
- Do not smoke in areas where cylinders are stored.
- Shade cylinders from the sun if they are temporarily stored outside during the summer.
- Do not store empty cylinders with full cylinders. Tear the tags provided with the cylinder so that it can be quickly ascertained if the tank is: full, in use, or empty.
- Label the storage room, and keep it cool, dry, and well ventilated.
- Do not let anything heavy fall on the cylinders.

Opening and Closing Valves

By observing a few simple rules when opening and closing valves, damage to valves and equipment can be prevented, adding years of useful service life to the valves.

The proper way to open any cylinder valve connected to a piece of equipment is to first crack the valve, then open it slowly by turning the handle or stem counterclockwise. This allows the equipment to gradually adjust to full pressure. Stop turning as soon as there is any resistance. Turning the valve handle or stem too far in the open position can cause damage and leaks to the cylinder, preventing proper closure. Likewise, over-tightening can damage or permanently distort the seat, resulting in leakage.

Specific Gases

Gases that are most commonly used in medical practices include nitrous oxide, carbon dioxide, medical air, oxygen, nitrogen and nitrogen NF, liquid nitrogen, and argon.

Nitrous Oxide

Nitrous oxide, or N_2O (ie, laughing gas, hippie crack), is a clear, colorless, oxidizing liquefied gas with a slightly sweet odor. The product is stable and inert at room temperature. While classified by the US DOT as a nonflammable gas, nitrous oxide supports combustion and can detonate at temperatures in excess of 650°C (1,202°F).

Nitrous oxide has beneficial use in a number of legitimate applications, particularly when blended with oxygen for use as an anesthetic. Workers can be exposed to N_2O while administering the anesthetic to patients, if fittings are not tight, or during recovery, when patients can exhale significant amounts of the gas. The effects of inhalation exposure to nitrous oxide include dyspnea, drowsiness, headaches, and asphyxia, at very high concentrations.

Standards and recommendations for the use of nitrous oxide include:

- OSHA Regulation (Standard Number 29 CFR 1910.105, *Nitrous Oxide*)
- NIOSH Alert, *Request for Assistance in Controlling Exposures to Nitrous Oxide during Anesthetic Administration*, Publication Number 94-100[1]
- NIOSH HC 29, *Control of Nitrous Oxide during Cryosurgery*[2]
- CGA SB-6—1999, *Nitrous Oxide Security and Control*[3]

The anesthetic effects of nitrous oxide begin to take effect at concentrations of about 10%. At higher concentrations, approaching 100%, a sense of well being or a high is experienced. A person experiencing a nitrous oxide high may have slurred speech, difficulty maintaining balance or walking, be slow to respond to questions, lapse into unconsciousness, or be immune to any stimulus, such as pain, loud noises, and speech.

Because of its relatively low toxicity and its anesthetic effects, nitrous oxide has gained popularity as a substance for recreational abuse. Members of the Compressed Gas Association (CGA) and of the National Welding Supply Association (NWSA) are joined by the medical profession, law enforcement, government, and the media in expressing concern regarding the abuse of nitrous oxide, which results in death and injury.

Those involved with nitrous oxide abuse often steal the cylinders from distributors or legitimate users or by illegally obtaining a legitimate nitrous oxide use permit.

Theft of cylinders makes nitrous oxide available to people seeking the euphoric qualities of the gas, but who are unaware of the hazards of abusing nitrous oxide. The CGA and NWSA have recommended guidelines for nitrous oxide sales and security. These guidelines are intended to help implement the principles of product stewardship, which include identifying sufficiently responsible control measures that minimize the theft of nitrous oxide and deter its abuse. Table 5.2 provides security recommended guidelines for medical practices.

To protect workers from the health risks associated with nitrous oxide, scavenging systems are sometimes used that vent unused and exhaled gas away from the work area. Recent research shows that these systems

TABLE 5.2

Nitrous Oxide Sales and Security Recommended Guidelines for Legitimate Users

Legitimate users—anyone who can substantiate a medical, industrial, food, or other legitimate use for nitrous oxide. It is recommended that legitimate users:

1. Minimize theft and indiscriminant use of nitrous oxide by storing containers and utilization equipment in a secured area subject to removal by authorized personnel only.
2. Keep an inventory of bulk product and/or containers (both full and empty) and investigate any discrepancies.
3. Promptly report any thefts to the police and shipper.
4. Use denatured (eg, rendered unfit for human consumption) nitrous oxide, if required, for enhancing performance of internal combustion engines, so as to deter abuse.
5. Alert employees to the dangers of nitrous oxide abuse and train them on the special security measures they should take to prevent its theft.
6. Provide employees with a current MSDS and any additional instructions.

Reproduced with permission from *Nitrous Oxide Sales and Security Recommended Guidelines*, published by the Compressed Gas Association. For more information visit www.cganet.com.

can significantly reduce the risk of impaired fertility among females exposed to nitrous oxide. However, simply using a scavenging system is not sufficient. It must be continuously monitored and maintained to effectively control exposure to nitrous oxide.

NIOSH provides information concerning the monitoring of anesthetic equipment to reduce nitrous oxide exposure in the workplace. Table 5.3 summarizes that information.

Carbon Dioxide

Carbon dioxide (carbonic acid gas) is a colorless, odorless, and slightly acid-tasting gas. Carbon dioxide does not burn or support ordinary combustion.

The presence of carbon dioxide in the blood stimulates breathing. For this reason, carbon dioxide is added to oxygen or ordinary air in artificial respiration and to the gases used in anesthesia for some minor surgical procedures. In a medical practice, carbon dioxide is also used with a CO_2 laser.

A carbon dioxide extinguisher, found in most medical practices, is a steel cylinder filled with liquid carbon dioxide, which, when released, rapidly expands. That rapid expansion causes a great lowering of temperature, resulting in solidification of the carbon dioxide into powdery

TABLE 5.3

Summary of Information Concerning the Monitoring of Anesthetic Equipment to Reduce Nitrous
Oxide Exposure in the Workplace

Monitor anesthetic equipment when installed and every 3 months thereafter:
- Leak test equipment
- Monitor air in the worker's personal breathing zone
- Monitor the environment (room air)
- Prevent leakage from the anesthetic delivery system through proper maintenance and inspection of equipment.

Eliminate or replace the following:
- Loose-fitting connections
- Loosely assembled or deformed slip joints and threaded connections
- Defective or worn seals, gaskets, breathing bags, and hoses

Control waste nitrous oxide with a well-designed scavenging system that includes the following:
- Securely fitting masks
- Sufficient flow rates (eg, 45 liters per minute) for the exhaust system
- Properly vented vacuum pumps

Make sure that the room ventilation effectively removes waste nitrous oxide. If concentrations of nitrous oxide are above 25 ppm, take the following steps:
- Increase air flow into the room
- Use supplemental local ventilation to capture nitrous oxide at the source

Institute an educational program that describes nitrous oxide hazards and defines prevention measures.

Courtesy National Institute for Occupational Safety and Health, *NIOSH Alert: Request for Assistance in Controlling Exposures to Nitrous Oxide During Anesthetic Administration*, DHHS (NIOSH) Publication "snow." This snow vaporizes on contact with the burning substance, producing a blanket of gas that eliminates oxygen and smothers the flame.

Medical Air

Air (compressed air, synthetic air, reconstituted air, or medical air) is a colorless, odorless, nonflammable gas. It is an oxidizing agent and may accelerate combustion. Contact with flammable materials may cause fire or explosion.

Medical air is often used to dry instruments, such as bronchoscopes and other items that have been treated with liquid disinfectants. Medical air can also be combined with pure oxygen to administer to patients.

Oxygen

Oxygen is a principal component of the air we breathe, but it exists at a concentration far lower than the 100% oxygen in a compressed gas cylinder. In a medical practice, patients who have lung and heart conditions use oxygen to increase the concentration of oxygen in lungs and blood.

The use of oxygen is addressed in several standards and guidelines, including:

- OSHA Regulation (Standard Number 29 CFR 1910.104, *Oxygen*)
- CGA SA-8—2000, *Hazards of Oxygen in the Health Care Environment*[4]
- CGA P-14, *Accident Prevention in Oxygen-Rich and Oxygen-Deficient Atmospheres*[5]
- NFPA 53, *Recommended Practice on Materials, Equipment, and Systems Used in Oxygen-Enriched Atmospheres*, 1999 ed. (1,2)[6]

Use oxygen with caution. Although oxygen itself does not burn, it does support and enhance combustion. Advances in material science have introduced more polymers (eg, endotracheal tubes) and fabrics (eg, drapes, gowns) that are normally not flammable at environmental concentrations of oxygen; however, they do become flammable in oxygen-rich atmospheres.

Similarly, materials that can be ignited in breathing air require less energy in an oxygen-rich atmosphere. Many of these materials can be ignited by the friction created by the opening of a valve seat, or by heat produced from adiabatic compression when oxygen at high pressure is rapidly introduced into a system of much lower pressure.

Some general guidelines when using oxygen and working in oxygen-rich atmospheres include:

- Do not allow valves, regulators, gages, or fittings to come into contact with oils, including skin oils, makeup, greases, organic lubricants, rubber, or any other combustible substance.
- Slowly and completely open oxygen valves to reduce heat production and achieve the desired flow within the equipment.
- Make sure that any cleaning or repair of oxygen-using equipment is performed by qualified, trained staff.
- Use plugs, caps, and plastic bags to protect "off-duty" equipment from dust and dirt.
- Make sure that personnel who operate oxygen-using equipment are adequately trained and understand the manufacturers' instructions for using the equipment.

Nitrogen and Nitrogen NF

Nitrogen is the largest volume inorganic chemical sold in the world, supporting a multitude of commercial and technical applications. It is nonflammable and is considered to be an inert gas (ie, it does not easily take part in chemical reactions). Nitrogen is colorless, odorless, and tasteless at normal temperature and pressure. There is no OSHA PEL or ACGIH TLV-TWA for this gas. It is listed as a simple asphyxiant.

In medical practices, nitrogen is sometimes used to power surgical equipment. The US Food and Drug Administration (FDA) does not specifically require that nitrogen, used to power surgical equipment, be listed as either a medical device or be nitrogen NF (ie, medical grade). Some laboratory equipment, especially equipment that might be used in a full CLIA laboratory, may be cleaned and calibrated using nitrogen gas.

To protect patients, the CGA recommends that manufacturers adopt a policy to only sell or distribute nitrogen NF for the purpose of powering surgical instruments and for any other applications in the health field where the end use application for the gas is unknown.

Nitrogen should be stored and used with adequate ventilation. The MSDS should be consulted for specific first aid, handling, storage, and exposure control information. Additional information is available from vendors and the CGA.

Liquid Nitrogen

In its liquid state, nitrogen is used for food freezing, plastic and rubber deflashing and grinding, cooling, metal treating, biological sample preservation, pulverization, and other temperature-related applications. It has no warning properties, and is inert, colorless, odorless, noncorrosive, and nonflammable.

A medical practice or clinic uses liquid nitrogen as a refrigerant to obtain and preserve biological specimens, such as tissues and growths, and in dermatologic practices for removal of skin disorders. Extensive tissue damage or burns can result from exposure to liquid nitrogen or cold nitrogen vapors. The main hazards associated with liquid nitrogen are:

- Burns, blisters, rashes, and deadening of the skin. Inhalation of the super-cold fumes can cause respiratory tract irritation and freezing.
- The displacement of oxygen in the air that we breathe. Because the gas expands very quickly and is produced in high concentrations from the cryogenic liquid, it displaces oxygen and a person may lose consciousness or die from the lack of oxygen.

Because of the large expansion ratio of liquid to gas, it is important to provide adequate ventilation, with a minimum of air changes per hour, in areas using liquid nitrogen. Oxygen concentrations should be monitored in areas where displacement may occur.

OSHA has established 19.5% oxygen concentration as the minimum level for working without supplied air. Personnel should not enter areas where the oxygen concentration is below this level unless provided with a self-contained breathing apparatus or air-line respirator.

Liquid nitrogen is usually stored in a special nonpressurized, thermos-type container called a Dewar bottle. The product may be removed from small Dewars by pouring, while larger liquid nitrogen transfers require a transfer tube.

> **BOX 5.1**
>
> **Exposure control methods for liquid nitrogen**
>
> The following includes special handling and control methods required for liquid nitrogen:
>
> **Storage**
>
> Liquid nitrogen should be stored in:
>
> - A locked room with a warning sign on the outside door (eg, Danger – Liquid Nitrogen. Freezing Hazard and Oxygen-Deficient Atmosphere Possible).
> - A room with adequate ventilation.
>
> **Handling**
>
> When handling liquid nitrogen, safety procedures recommend:
>
> - Use with adequate ventilation.
> - Do not allow the fumes or liquid to come into contact with electrical circuits.
> - Do not allow the liquid to come into contact with tissue, bone, plants, flesh, rubber, latex, plastic, wood, or meat. These items will immediately freeze and be subject to extreme damage.
> - Do not mix water, alcohol, or solvents with the liquid nitrogen.
>
> **Disposal**
>
> Dispose of small quantities of liquid nitrogen by letting the liquid warm to room temperature in a well-ventilated space. Use cold water for rinsing.
>
> **Personal Protective Equipment**
>
> Eyes are most sensitive to the extreme cold of liquid nitrogen. The recommended PPE when handling or using liquid nitrogen is a full face shield over safety glasses, thermal insulated or leather gloves, and long-sleeve shirts and trousers without cuffs, especially whenever the possibility of exposure or a spill exists. In addition, when handling liquid nitrogen containers, safety shoes are recommended. In emergency situations involving the breathing area, use self-contained breathing apparatus (SCBA). First-aid treatment should be the same as when treating severe frostbite.

Argon

Argon is nontoxic and odorless. Liquid argon is an extremely cold liquid and gas under pressure. Argon can act as a simple asphyxiant by displacing air or liquefying oxygen. In a medical practice, argon is used as a specialty gas for gas lasers and other equipment. Exposures to high concentrations of argon can lead to oxygen displacement and asphyxiation. Moderate concentrations may cause headache, drowsiness, dizziness, excitation, excess salivation, vomiting, and unconsciousness.

To avoid the harmful effect of argon, workers should follow strict safety guidelines for storage and handling and consult the MSDS for precautions and first-aid measures.

Odorizing Atmospheric Gases

Atmospheric gases, such as those described in this section, do not have any odor. Thus their presence cannot be detected by smelling. In some cases, such as natural gas used for home heating and cooking, an odorant is added to warn people of the presence of that highly flammable gas. However, the CGA does not recommend the odorization of atmospheric gases, of which the primary reasons are provided in Table 5.4.

SERVICING MACHINES AND EQUIPMENT (LOCKOUT/TAGOUT)

While energy runs industrial machines and equipment, it can be dangerous if not carefully controlled during the servicing or maintenance of machines and equipment. The failure to control potentially hazardous energy during routine service and maintenance procedures could present

TABLE 5.4
Primary Reasons Offered by CGA for Not Recommending the Odorization of Atmospheric Gases

1. The presence of an atmospheric gas does not create a hazard, the degree of concentration is the hazard. Degree of concentration cannot be determined by an odorant.
2. Odorization of atmospheric gases is not considered a substitute for good engineering practice, proper operational and maintenance procedures, and adequate training of personnel in understanding the hazards of oxygen-enriched, oxygen-deficient, flammable, and toxic gas atmosphere.
3. OSHA Standards 29 CFR 1910.146, Permit-Required Confined Spaces, and 29 CFR 1915, Confined and Enclosed Spaces and Other Dangerous Atmospheres in Shipyard Employment, contain requirements for practices and procedures to protect employees from the hazards of entry into confined spaces and activities conducted within the spaces. Some of the requirements of these standards are:

    - Continuously monitoring the space to ensure that the oxygen content is within 19.5-2 and to detect the presence of other toxic or flammable gases
    - Providing an uninterrupted supply of fresh air into the space
    - Establishing a hazardous work permit procedure for planning and executing the work activity, addressing such issues as safety watch and rescue requirements

4. Each gas would have to have its own unique odor. For example, an oxygen-enriched space would require an odor different from that used for an oxygen-deficient atmosphere created by an inert gas and felt to be a greater hazard in a confined work area.
5. The odorization of gases could create a "no smell, no hazard" assumption. For example, an oxygen-deficient atmosphere can be inside a closed vessel by the formation of oxidation products (eg, rust) or by the physical absorption of the oxygen.
6. Other smells and fumes generated in the work environment could mask the presence of other odorants. It is well known that a percentage of the population has a complete or partial loss of smell and that some people are insensitive to various odors.
7. CGA Position Statement, PS-3, on Odor Testing for Cylinder Contaminants (1990, reaffirmed 1996) does not recommend sniff testing of cylinders prior to filing because of potential exposure to hazardous contaminants. Knowingly odorizing a gas to indicate its presence is philosophically contrary to practices that are designed to reduce a worker's exposure to hazardous chemicals.

Reproduced with permission from CGA *Position Statement on Odorizing Atmospheric Gases (Oxygen, Nitrogen, and Argon)*, published by the Compressed Gas Association. Copyright 1996, Compressed Gas Association.

serious risks to workers and the medical practice, including human pain and suffering, dangerous exposure to other employees, and financial exposure from fines or accidents.

OSHA has issued a regulation called *The Control of Hazardous Energy (Lockout/Tagout)* (Standard Number 29 CFR 1910.147, Subpart J), which covers procedures to prevent unexpected energization, start-up, or release of stored energy that could cause injury to employees. The program procedures must include preparation for shutdown, equipment isolation, lockout/tagout application, release of stored energy, and verification of isolation. The standard also requires training for authorized and affected employees. Table 5.5 provides OSHA's definition of lockout and tagout.

Some medical practices use equipment that requires lockout or tagout before servicing or repair. In this case, the medical practice must develop and implement a program that complies with OSHA's standard.

If the medical practice repairs or services equipment in-house, the requirements are important to three groups of workers:

- Employees who lockout or tagout machines or equipment in order to perform servicing or maintenance.
- Employees who work with the equipment to be locked out or tagged out.
- Employees who work in the area in which servicing or maintenance is being performed.

When outside contractors provide servicing or maintenance work on the equipment, the medical practice and contractor must inform each other of their respective lockout/tagout procedures. Employees of the medical practice must understand and comply with the contractor's energy control procedures.

TABLE 5.5

Definitions of Lockout and Tagout

OSHA standards provide the following definition of Lockout:

Lockout. The placement of a lockout device on an energy-isolating device, in accordance with an established procedure, ensuring that the energy-isolating device and the equipment being controlled cannot be operated until the lockout device is removed

Lockout device. A device that utilizes a positive means such as a lock, either key or combination type, to hold an energy-isolating device in the safe position and prevent the energizing of a machine or equipment. Included are blank flanges and bolted slip blinds.

OSHA standards provide the following definition of Tagout:

Tagout. The placement of a tagout device on an energy-isolating device, in accordance with an established procedure, to indicate that the energy-isolating device and the equipment being controlled may not be operated until the tagout device is removed.

Tagout device. A prominent warning device, such as a tag and a means of attachment, that can be securely fastened to an energy-isolating device in accordance with an established procedure, to indicate that the energy-isolating device and the equipment being controlled may not be operated until the tagout device is removed.

Courtesy Occupational Safety and Health Administration, OSHA Regulation (Standard Number 29 CFR 1910.147, The control of hazardous energy).

SLIPS, TRIPS, AND FALLS

Slips, trips, and falls are among the most common workplace injuries in general industry, as well as medical offices. Many victims slip, trip, or fall on the same level where they are walking. Some of the most frequently overlooked hazards involve poor housekeeping. Preventing slips, trips, and falls requires a team effort to identify potential hazards and take corrective action before an injury occurs. Although slips and trips often result in falls, it is not necessary to fall in order to sustain an injury. The following sections are important to help prevent slips, trips, and falls in medical practice.

Slips

In order to prevent slips in the workplace, it is important to:

- Promptly clean up spills.
- Post signs to warn of wet areas.
- Have ice and snow cleared as quickly as possible.
- Shelter doorways from ice, snow, sleet, and rain.
- Have slip-resistant coatings or products installed in areas likely to become slippery and on stair treads.
- Be careful of wet shoes on a dry floor. They can be just as slippery as dry shoes on a wet floor.

Trips

In an effort to prevent trips in the workplace, employees of the medical practice should:

- Turn on lights before entering an area and have light bulbs replaced when burned out.
- Use footstools with rubber feet.
- Use stepstools for reaching high places.
- See that damaged sidewalks, parking areas, and other walking surfaces are repaired as quickly as possible.
- Have loose or broken grab bars and handrails repaired.
- Have loose carpet tacked or taped down.
- Use throw rugs with a skid-resistant backing.
- Cover cables that cross walkways to reduce trips.
- Keep walkways free of objects and clutter.
- Put trash in the trash cans.
- Close file cabinet drawers.

Falls

By observing the following safety precautions, most falls can be avoided:

- Keep chairs, exam tables, and other equipment in good repair.
- Use ladders properly.

- Take steps one at a time and use the handrail. Never skip or jump from one level to another.
- Never use the stairs as a storage room.
- Slow down when approaching steps or a staircase.

An effective prevention program focuses on the potential for these hazards. The following should be included in the slip, trip, and fall prevention program:

- Training in the identification of potential slip, trip, and fall hazards.
- Immediate reporting of any potential physical hazards due to slips, trips, and falls including those reported by patients.
- Records of incidents and information regarding slips, trips, and falls.
- Routine inspections of the facility for possible hazards.
- Provisions for working with the contractor and architect when remodeling and building new facilities.

An effective and well-managed prevention program can reduce liability and increase staff, patient, and visitor safety and well-being.

ENDNOTES

1. National Institute for Occupational Safety and Health (1994). NIOSH Alert, *Request for Assistance in Controlling Exposures to Nitrous Oxide during Anesthetic Administration*. US Department of Health, Education, and Welfare, Public Health Service, Centers for Disease Control and Prevention, National Institute for Occupational Safety and Health. DHHS (NIOSH) Publication No. 94-100.
2. National Institute for Occupational Safety and Health (1999). NIOSH Hazard Controls, *Control of Nitrous Oxide during Cryosurgery*. US Department of Health, Education, and Welfare, Public Health Service, Centers for Disease Control and Prevention, National Institute for Occupational Safety and Health. DHHS (NIOSH) Publication No. 99-105.
3. Compressed Gas Association (1999). *Nitrous Oxide Security and Control*. Safety Bulletin SB-6—1999.
4. Compressed Gas Association (2000). *Hazards of Oxygen in the Health Care Environment*. Safety Alert SA-8—2000.
5. Compressed Gas Association (P-14). *Accident Prevention in Oxygen-Rich and Oxygen-Deficient Atmospheres*.
6. National Fire Protection Association 53, *Recommended Practice on Materials, Equipment, and Systems Used in Oxygen-Enriched Atmospheres*. 1999.

chapter 6

Medical Equipment Safety Management

Proper maintenance management ensures that medical equipment is safe, accurate, and ready for use when needed at the lowest possible cost.
Al Kuntz, Vice President, Asset Technology & Facilities Management Professional Services

Medical practices have legal duties to provide safe patient care equipment. In some medical practices, the safety responsibility for this equipment is shared with either a hospital biomedical department or an outside service organization. However, the medical practice is ultimately responsible for ensuring that equipment used in the practice will perform as expected and present no risks to patients or staff.

Careful equipment management offers the potential for reducing liability and risk of injury to patients and employees and lessens the chance of significant adverse effects on patient care.

STANDARDS AND RECOMMENDATIONS

Certain organizations and government bodies have developed guidelines and standards that address the maintenance of medical equipment. Many of these guidelines and standards are also referred to as "industry standards" and may be used in a court of law during a case that involves injury or death caused by malfunctioning medical equipment.

Safe Medical Device Act

The Safe Medical Device Act (SMDA) of 1990 requires health care facilities to report all medical-device-related serious injuries and deaths to the manufacturer. The FDA must be notified of all incidents that result in death. OSHA must be notified if the incident results in the death of a staff member and/or the hospitalization of three or more staff members. The SMDA gives FDA investigators access to the facility and to organizational equipment records.

In an effort to reduce medical device malfunctions and mishaps, medical practices should:

- Develop a medical device/equipment management program.
- Establish formal reporting and investigation procedures.

- Train staff members how to respond to potentially serious medical device/equipment incidents.
- Never use or release a medical device, accessory, or piece of equipment involved in a reportable incident to anyone until independent testing has been conducted and documented.

Staff at medical practices, especially practices that house and use medical devices and equipment, such as lasers, X ray equipment, and electrocautery units, should set up a medical equipment management program that includes the tenants of the SMDA.

OSHA General Industry Standards

There is no OSHA standard that specifically addresses medical equipment hazards. The enforcement of the medical equipment safety program falls under the OSHA General Duty Clause (5a1) and the OSHA Regulation, Subpart S—*Electrical* (Standard Number 29 CFR 1910.301 to 399).

Laser Standards, Resources, and Recommendations

There are a variety of laser safety standards including federal and state regulations and nonregulatory standards. The most often quoted is the American National Standards Institute's Z136 series of laser safety standards. These standards are the foundation of laser safety programs in industry, medicine, research, and government. The ANSI Z136 series of laser safety standards are referenced by OSHA as the basis of evaluating laser-related occupational safety issues.

ANSI Z136.1, *Safe Use of Lasers*,[1] is the parent document in the Z136 series and provides information on laser safety and hazard control measures. This document serves as a reference for laser safety officers and laser safety committees.

The Center for Devices and Radiological Health (CDRH) and the FDA regulate product performance, and all laser products sold in the United States since August 1976 must be certified by the manufacturer as meeting certain product performance (safety) standards. Each laser must bear a label indicating compliance with the standard and denoting the laser hazard classification.

Additional laser resources include the following:

- OSHA Regulation, Subpart I, Personal Protective Equipment (Standard Number CFR 1910.132, *General Requirements*) for face and eye protection.
- ANSI Standard Z136.3, *Safe Use of Lasers in Healthcare Facilities*, provides guidance for laser use in medicine.[2]
- Food and Drug Administration (Standard Number 21 CFR 1040.10 – 11, *Performance Standards for Light-Emitting Products*).[3]
- ACGIH recommendations to reduce occupational exposure to laser hazards.[4]
- Conferences, laser safety courses, and safety bulletins from the Laser Institute of America.
- NFPA 99: *Health Care Facilities Handbook, Annex 2*, contains safety information for laser use in health care facilities.[5]

Joint Commission on Accreditation of Healthcare Organizations

Hospitals and certain other health care organizations are required to develop and implement plans for the management of medical equipment under standards issued by the Joint Commission on Accreditation of Healthcare Organizations (JCAHO). Specifically, the commission's standards for management of the environment of care (EC) require health care organizations to develop a written plan that describes how they will establish and maintain a management program to promote safe and effective use of medical equipment. Health care facility plans must include provisions for selecting, acquiring, and keeping track of medical equipment, as well as evaluating and minimizing risks associated with the equipment.

The EC Medical Equipment Standard also requires that the plans address equipment hazard notices and recalls, monitoring and reporting device-related injuries or deaths under the SMDA, and reporting and investigation of equipment management problems, failures, and user errors. In addition, the plans must provide for medical equipment orientation and education for the staff, and performance standards must be developed to evaluate and improve the effectiveness of the medical equipment program.

The JCAHO usually views the equipment management program as a four-step process, which includes:

1. Develop a policy to meet the organization's goals and objectives.
2. Establish equipment performance monitoring criteria and procedures.
3. Collect and document equipment performance data.
4. Analyze data to identify problems and ways to strengthen the program.

Although most medical practices do not fall under the auspices of JCAHO, it would be prudent to become familiar with the requirements of JCAHO's Medical Equipment Standard before developing a program.

Other Agencies

Individual states may have further requirements for medical equipment safety management programs and reporting requirements under the state department of health, local fire department, or other entities. It is recommended that medical practices check with the state and local governing bodies before they write programs, policies, and procedures.

TYPES OF MEDICAL EQUIPMENT/DEVICES TYPICALLY FOUND IN MEDICAL PRACTICES

Medical practices typically have on site the medical equipment listed below:

- Sterilizer (usually a table-top steam model, but could also have a chemical sterilizer)
- Crash cart with defibrillator
- Exam tables (can be electrical, manual, or stationary)
- Exam lights

- Otoscopes (electric power source)
- Ophthalmoscopes (electric power source)
- Pulse oximeters
- Ventilators
- Lasers
- EKG monitors
- EEG machines
- Anesthesia machines
- X ray equipment
- Centrifuges
- Nebulizers
- Ultrasound equipment
- Cast saws
- Incubators
- Eye exam chairs
- IV pumps
- Suction apparatus
- Air/oxygen regulators and flowmeters
- High-pressure supply tanks for oxygen and other compressed gases
- Other monitoring equipment (eg, treadmill monitors)
- Treadmills
- Exercise bikes
- Other electronic, electrical, pneumatic, manual, or hydraulic diagnostic, positioning, or treatment equipment/devices

PROCUREMENT PLANNING

Effective procurement planning enables medical practices to select devices that, in addition to being safe and productive, offer the best combination of ease of use, effortlessness of servicing, and reliability performance with the least amount of preventive maintenance. User training programs and operation and repair documentation, provided by the manufacturer, should also be investigated before purchasing. Once equipment is purchased, thorough acceptance testing should be conducted to reveal unexpected problems in manufacture or design. Acceptance testing may also be required following major repairs or software upgrades. A good procurement planning program includes all facets of the equipment life cycle and determines not only the equipment serviceability, but the total cost of ownership throughout time. Table 6.1 provides tips on purchasing equipment.

ELEMENTS OF A MEDICAL EQUIPMENT SAFETY PROGRAM

All patient care equipment used in a medical practice should be included in an organized program. Rarely is staff directly responsible for inspecting or maintaining medical equipment. Typically, a contracted service company performs that function.

A comprehensive program also addresses preventive maintenance. Equipment that is frequently used (eg, medical exam tables, ophthalmo-

TABLE 6.1

Tips for Purchasing Capital Patient Care Equipment

1. *Plan for what the medical practice needs.* The more thorough in this first phase of the process, the more success with the next steps and the happier staff will be with the equipment. Be realistic about what the office actually needs (as opposed to what the staff might want). Carefully analyze the different variables, such as volume, capacity, proprietary consumables, software and hardware upgrade ability, and required preventive maintenance. Consult available data, and then compare potential purchases with the medical practice's big picture. Use a consultant if necessary.

2. *Shop wisely.* The acquisition phase also requires planning. Research vendors and compare sales literature and information found in professional journals and on the Internet. Attend trade shows, talk to the reps, and check with peers. Be inspired, but do not let all the glitzy giveaways blind a decision. Do not just choose equipment; choose a company, remembering that you purchase a *capability,* not just a device. Learn about its business practices through shareholders reports, customer service, research and development, and expansion. Climb the corporate ladder to get answers to questions. In the end, choose value over low price, or the practice risks getting what they paid for.

3. *Consider life cycle issues.* The initial cost of equipment may be less than half the cost of ownership. Be certain to include space, staffing and training requirements, installation costs, lifelong maintenance, and proprietary consumable costs to determine the true cost of ownership. Explore what products or assistance an equipment supplier might be able to provide at low cost to them, but at great value to the practice.

4. *Implement in reverse.* Determine the "go-live" date, and work backward to see what must be accomplished when. Establish an implementation team to oversee all issues, including staff training, equipment calibration, and controls. After implementation, track quality, reliability, and staff productivity for use in the practice's next capital acquisition.

Adapted with permission from the Institute of Management and Administration. Copyright IOMA's Medical Laboratory Management Report. Sherrye Henry Jr, Editor. (212) 241-0360. For more information visit www.ioma.com.

scopes, X-ray equipment) in a harsh environment (eg, high humidity, heat), or is wholly or partially a mechanical assembly (eg, rehabilitation equipment), is likely to require preventive maintenance. An equipment management program that identifies, maintains, and improves performance and safety will help detect equipment damage caused by abuse, neglect, age, or accidents.

Each medical practice has different kinds of equipment. As old equipment is discarded and new purchased (or donated to the practice), the safety program should be changed. Review the program and revise it annually or any time the equipment changes. The medical equipment safety program should be easily accessible and readily available to all staff, who should receive training in the content of the program. The medical equipment safety program is aimed at protecting both the staff and the patients.

All medical equipment, no matter how new or old, should have an owner's and technical manual. File the original and attach a copy of the manual to or in close proximity to its corresponding piece of equipment. If there is no manual available, contact the manufacturer.

Risk Assessment

The first step in planning for the medical equipment safety program is an analysis of risks that are presented by such equipment in both the patient care and staff safety processes. The risks can range from life threatening (eg, the equipment short circuits and the patient or staff member is electrocuted) to procedure threatening (eg, the equipment needs to be repaired and cannot be used for specific treatments).

An effective risk analysis should address engineering and the medical equipment. An engineering risk analysis should include:

- *Physical risks.* For example, equipment located in a wet environment would present greater risks than when operated in a dry environment.
- *Maintenance requirements.* Devices with direct patient connections or sophisticated electromechanical features will likely require more extensive maintenance than simple equipment that never comes in contact with the patient.
- *Incident history with individual types or pieces of equipment.* Through experience or consultation with other equipment users, it may be discovered that certain types or pieces of equipment are more maintenance intensive than others or more troublesome and represent increased risk of failure during use in a medical practice.

A medical equipment risk analysis should:

- Identify the types of patient care equipment in the facility.
- Define the engineering and other potential problems associated with each type or piece of equipment (eg, calibration, the need for preventative maintenance).
- Identify competence factors necessary for proper equipment operation, including staff knowledge of instructions for setting up and operating the equipment, as well as interpreting and using the information provided by the equipment manufacturer.
- Identify the direct effect of equipment operation on patient health. Address the impact of equipment malfunction, ranging from minor nuisance to life-threatening situations.
- Identify how equipment will be allocated for use. The facility must be prepared to meet community and patient medical needs.

Inventory of Needs

After completing the risk assessment, develop an inventory of medical equipment included in the medical practice management program. This inventory should address all of the factors covered by the risk assessment.

Specifically, the inventory should list the various types of equipment and related requirements for engineering and educational support, as well as competence measurement, as appropriate. Identify resources needed to carry out the medical equipment management program and list staff who may become involved in the process.

Basic Program Components

The medical equipment management program should consist of a management plan supported by operating procedures. The plan describes how to manage the equipment and states who has the authority and responsibility for successfully carrying out the program.

The procedures describe in detail which specific activities must be performed (eg, daily safety checks, electrical inspections, calibration) and when they are to be performed in order to manage the risks that are associated with each specific piece of equipment. Only address and include activities that are potentially affecting risk (eg, electrocution, incorrect dosage).

The Management Plan

The medical equipment management plan should include the following:

- *Purpose statement defining the scope of the program.* It generally addresses equipment evaluation and acquisition, education of operators and users, maintenance and testing, and performance measurement.
- *A description of the program activities.* These activities include management responsibilities, such as policy and procedure development, training and education, equipment acquisition, maintenance, repair, and performance.
- *Documentation system that shows activities that are to be tracked and documented.* This system should support timely performance of the required tasks and a means by which such performance can be documented in case a device is involved in a negative patient outcome.

Program Procedures

Activities forming the basis of medical equipment management can generally be grouped into three types of procedures: technical testing and maintenance, administrative, and emergency and incident responses. The more medical equipment in the facility, the larger the facility, the more complicated the equipment, the more aged the medical equipment, and the more operators and users, the more complicated the procedures must be. The following are the three main types of procedures found in medical equipment management:

- Technical testing and maintenance procedures are usually developed from manufacturer recommendations, experience, and expert resources. Testing and maintenance schedules, as well as the content of testing and maintenance procedures, may be very flexible, as long as they are based on risk analysis and actual experience with the equipment. Document all analysis, results, repairs, and protocols, and keep them on site.
- Administrative procedures delineate the relationships among persons responsible for carrying out the program and describe activity schedules, work methods, reporting procedures, and communication formats and schedules.
- Emergency response procedures address patient risk associated with the disruption or failure of a piece of medical equipment (eg, a defibrillator). Other disruptive factors such as natural and human-caused disasters should also be addressed. The procedures must describe the essential communications that are needed to initiate an appropriate

clinical response, acquire backup equipment, and obtain emergency repair service. Incident response procedures direct staff to the appropriate actions in the event of a negative patient outcome—regardless of severity. The layout of circumstances, including equipment setup and configuration, must be maintained until a proper investigation and documentation can be accomplished.

Training and Education

The training and education process is usually comprised of four components: education, training, orientation, and competence. Do not consider these components separate entities, but rather as parts of a continuous process of interrelated, ongoing activities designed to establish and reinforce staff medical equipment knowledge. The following bulleted list outlines these four essential components:

- *Education* is the development of new knowledge through formal classroom or mentoring processes or by hiring or credentialing new staff members. The goal of medical equipment education is to expand possible treatment alternatives that are available to the office, practice, or clinic. This new knowledge slowly develops and requires experience to become useful.
- *Training* is designed to quickly develop mechanical or repetitive skills. Training involves teaching staff operators/users how to set up, adjust, and analyze the performance of the medical equipment. The goal is to minimize a patient's risk of injury due to errors (eg, electrical shock, misdiagnosis).
- *Orientation* is the process of teaching new staff about organizational expectations to ensure that work processes are consistently carried out.
- *Competence* can be characterized as the ability to perform all the mental and physical parameters of a given job or profession. These include mechanical skills, as well as cognitive ability and wisdom.

Program Implementation

Once the medical equipment safety management program is developed, it needs to be implemented as designed to keep it running efficiently. As adjustments are made, program documentation must be updated to reflect new policies and procedures.

Program Monitoring and Measurement

Monitoring and measuring of the medical equipment management program is necessary to evaluate various aspects of the program, ensure program effectiveness, and enable performance improvement. Program aspects to measure include:

- The management schedule to ensure that all relevant schedules are followed. This should cover equipment acquisition, maintenance, and repair, as well as staff orientation, training, and competence measurement.

- Measurement of staff performance in terms of equipment use, operation, and maintenance.
- Measurement of equipment reliability and effectiveness. To measure this reliability, monitor a specific, relevant attribute of the specific piece of equipment, such as mean time between failures or breakdowns per number of uses.

Data Evaluation and Communication

As the various aspects of a medical equipment management program are measured, the results must be periodically evaluated and communicated, as appropriate, to maintain and improve program effectiveness. Some of the steps involved in this process include:

- *Data selection and definition.* Data (eg, operability of the piece of equipment/device) gathered on the medical equipment management program should directly relate to daily work activities and be collected as a part of routine work.
- *Gathering data.* Gather data in a uniform and timely manner. Follow a set schedule to ensure a continual stream of data and to guarantee that adequate information from the appropriate, involved personnel who actually use, operate, and maintain the medical equipment/device is secured.
- *Forming information.* Once gathered, assemble the data into useful information (eg, graphs, charts, written summaries) that gives the medical equipment users, operators, and maintainers the information necessary to ensure that all medical equipment and devices are fully operational and as safe as possible.
- *Applying wisdom.* When problems are identified, the medical office, practice, or clinic management must exercise wisdom to balance both financial considerations and resource limitations with the need to make changes and replace or repair equipment.

Program Improvement

Improvement of the medical equipment management program can be achieved in many ways. The most important steps in the program improvement process include:

- Defining the problem.
- Forming and testing a hypothesis for improvement.
- Evaluating results.
- Retesting the hypothesis.
- Incorporating the resulting solution into everyday use.

Malfunctioning medical equipment or devices can endanger both staff and patients. Developing and implementing a strong medical equipment management program, based on standards and recommendations from OSHA, JCAHO, SMDA, and the FDA, can help ensure the safety and health of all involved.

LASERS

Laser (light amplification by stimulated emission of radiation) use is increasing at a very fast pace in the health care environment, including medical practices. New laser surgery techniques are being developed almost daily and can be used for medical procedures, from corrective eye surgery to helping align patients for treatments in radiology departments.

Classes of Lasers

There are four FDA laser classes:

- Class 1. Lasers capable of producing damaging radiation.
- Class 2. Lasers that may be viewed under strict controls.
- Class 3. Lasers that may require controls to prevent direct view and eye damage.
- Class 4. Lasers that must be controlled to prevent eye and skin damage (medical use lasers usually fall into this class).

Laser Beam Hazards
The laser produces an intense, highly directional beam of light. The primary safety concern with lasers is the possibility of eye damage from exposure to the laser beam during corrective eye surgery. A thermal burn destroys the retinal tissue. Acoustic damage usually affects a greater area of the retina and is more destructive than a thermal burn. Since retinal tissue does not regenerate, the damage is permanent. The nature of the eye damage depends on the type, power, and duration of the laser exposure. In other treatments, lasers striking the skin can result in erythema, blistering, and charring. The extent of the damage in these instances depends on wavelength, power, and length of exposure. Lasers also use high voltage and can be considered a potential electrical hazard.

Laser Nonbeam Hazards
In addition to the direct hazards to the eye and skin from the laser beam itself, there are other hazards associated with the use of lasers. These nonbeam hazards, in some cases, can be life threatening (eg, electrocution, fire, and asphyxiation). Fire is also a significant hazard associated with lasers. All combustible materials within the area of laser use are susceptible to ignition in the event the laser beam is directed to it. For this reason, it is imperative that all combustible materials, including drapes and paperwork, be carefully managed during laser procedures.

Laser Safety Guidelines
The following include laser safety guidelines:

- Use effective eye protection.
- Appoint a laser control officer.
- Only trained and authorized technicians should move, operate, or repair laser equipment.
- Attach lasers to an individual transformer and an emergency power source with a safety interlock.

- Keep an approved fire extinguisher immediately available.
- Identify laser use areas and post laser warning signs.
- Prevent laser beams from coming into contact with combustible, flammable, and reflective materials.
- Personnel using or being exposed to lasers should be in an eye health medical surveillance program.
- Promptly remove and filter smoke generated during laser surgery before evacuating it from the building.
- Properly maintain laser equipment and include it in a preventive maintenance program.
- Report laser injuries through the FDA's SMDA.

ENDNOTES

1. American National Standards Institute (2000). Z136.1 *Safe Use of Lasers* (2000 ed.). Laser Institute of America.
2. American National Standards Institute (1996) Standard Z136.3, *Safe Use of Lasers in Healthcare Facilities*. Laser Institute of America.
3. US Food and Drug Administration (2000). *Performance Standards for Light Emitting Products*. (Standard Number 21 CFR 1040). Revised April 1, 2000. Title 21.
4. American Conference of Governmental Industrial Hygienists, Inc; Cincinnati, OH.
5. National Fire Protection Association (1999). *NFPA 99: Health Care Facilities Handbook, Annex 2*.

chapter 7

Personal Protective Equipment

Failure to wear PPE is like bungee jumping without a cord.
Yvonne MacManus, Chaff & Co

Accidents can happen in just a second. Countless reports tell about employees who removed their protective eyewear for just a moment and liquid chemicals spattered into their eyes or employees who performed a quick task involving blood or body fluids without putting on the necessary gown and their clothes and skin were contaminated, resulting in dangerous health risks. Lack of attention is a frequent cause of minor and serious injuries and more often occur when employees are distracted by external factors, feel pressured by circumstances that get in the way of safely performing the job, or do not believe safety is an important part of their employer's culture.

Personal protective equipment (PPE) is an essential part of many jobs. PPE may be required to reduce an employee's risk of exposure by contact, inhalation, or ingestion of an infectious agent, toxic substance, or radioactive material. PPE protects employees from injury by creating a barrier against workplace hazards. It is not a substitute for good engineering, administrative controls, or good work practices, but should be used with these controls to ensure the safety and health of employees.

In medical practices the most common types of PPE are gloves, gowns, protective eyewear, fluid-resistant masks, respiratory devices, and face shields. Not everyone needs them, so OSHA requires employers to determine what is needed based on an assessment of workplace hazards.

PROGRAM COMPONENTS

Following are program components to OSHA's Regulation (Standard Number 29 CFR 1910.132–139, *Personal Protective Equipment*):

- Hazard assessment and protective equipment selection
- Protective devices
- Cleaning and maintenance
- Employee training
- Recordkeeping

Hazard Assessment and Protective Equipment Selection

The first step in providing a safe work environment is assessing the hazards that employees may encounter and determining what kind of PPE is needed. This process is very similar to the Hazard Communication Standard (HCS) discussed in chapter 4. Under the HCS, employees must be told about any hazardous conditions they may find or any hazardous materials with which they may come into contact so that they may better protect themselves.

Employees should be involved in the hazard assessment. They are the ones doing the work and know what is involved in their tasks. Employee involvement is successful in work environments that actively encourage the staff to speak up and make suggestions.

In medical practices, hazards are influenced by the type of medicine practiced (eg, orthopedic, gastroenterology, pulmonary), as well as the in-house services, such as a laboratory. Hazards may include sterilants and disinfectants, contaminated laundry, infectious diseases, medical waste, or chemical, biological, and radiological spills. Table 7.1 provides hazard assessment guidelines as summarized by OSHA's Standard for Personal Protective Equipment (Standard Number 29 CFR 1910.132).

Employers must certify in writing that a workplace hazard assessment has been performed. The document must include date(s) of the assessment, the name of the person who performed the assessment, the findings of the assessment, and specific recommendations for protective equipment needed. The assessment form, a sample of which is provided in Figure 7.1, should be kept for 3 years.

After identifying the hazards in the workplace, determine the suitability of the PPE that is presently available. *Defective or damaged personal protective equipment must not be used.* If necessary, select new or additional

TABLE 7.1

Hazard Assessment Guidelines

1. *Conduct a Hazard Assessment.* Inspect the medical practice for potential hazards that include: equipment that could be injurious if improperly used or through equipment failure; blood or body fluid splashes; chemical exposure; high temperatures that could lead to eye injury, burn, or fire; harmful dusts; sources for light (optical) radiation; potential sources for falling objects; procedures involving sharps; and possible electrical dangers.

2. *Document hazards.* During the hazard assessment, potential causes—from fixtures to equipment to procedures—should be written down immediately to avoid overlooking anything. Organize the information by the nature of the hazard (eg, eyes, hands) to establish what PPE may be needed.

3. *Analyze the potential for injuries.* Determine the types of hazards, level of risk, and seriousness of potential injury from each hazard found. In a medical practice, employees might be exposed to several hazards simultaneously, and this should be part of the analysis. Isolate the type(s) of potential injury to determine what type(s) of PPE is necessary.

4. *Select PPE that is task specific.* To ensure employees are protected, PPE should provide a level of protection greater than the minimum required. Each employee required to wear PPE should be fitted with the right protection. PPE must fit comfortably and securely to provide maximum protection. Instruction must be provided on the care and use of PPE

5. *Reassess workplace hazard potential as necessary.* Identify and evaluate new equipment and procedures, review accident records, and compare the suitability of current PPE against newer, safer types of PPE.

Courtesy Occupational Safety and Health Administration, OSHA Regulations (Standard Number 29 CFR 1910.132).

FIGURE 7.1
A Sample of a Certification of PPE Hazard Assessment Form
Reproduced with permission from McFadden & Associates.

Certification of PPE Hazard Assessment

Surveyor: _____   **Date:** _____

Office/Clinic/Practice: _____   **Phone:** _____

Per OSHA 29 CFR 1910.132, paragraphs (d), (f), and (g), Personal Protective Equipment, General Requirements, the PPE hazard assessment and equipment selection does <u>not</u> apply to .134 (respirators),.137 (electrical protective devices) and .139 (respiratory protection for M. tuberculosis). 1910 Subpart I, Appendix B, is <u>non-mandatory</u>, and addresses the compliance guidelines for hazard assessment and personal; protective equipment selection (use this appendix as a guide).

❏ **NO ASSESSMENT REQUIRED** — No hazards exist that require the use of PPE, or are likely to require the use of PPE

POTENTIAL HAZARD SOURCE	ASSESSMENT OF HAZARD	PPE REQUIRED
Use or handling of: ❏ Chemicals ❏ Biologicals ❏ Radioactive materials	Possible injuries: ❏ Impact from flying particles ❏ Chemical splash in the eyes ❏ Facial/skin chemical contact ❏ Biological contact with mucous membrane ❏ Body/skin/hand contact with biological agents, sharps, radioactive materials, chemicals, or hot or cold objects	Types of PPE required for protection: ❏ Safety glasses with side shields ❏ Safety goggles ❏ Chemical splash goggles ❏ Face shield ❏ Fluid resistant face mask for biologicals ❏ Respirator for TB and other airborne pathogens* ❏ Fluid resistant lab coat ❏ Plastic/rubber apron ❏ Fluid resistant gowns ❏ Chemical resistant suits/gowns ❏ Chemical resistant gloves ❏ Body fluid resistant gloves ❏ Non-latex gloves ❏ Chemical and fluid resistant boots ❏ Lead aprons ❏ Other
❏ High noise levels from equipment	❏ Sound levels in excess of OSHA PEL ❏ Staff hearing loss	❏ Ear muffs ❏ Ear plugs — disposable ❏ Earplugs — reusable ❏ Other
Non-ionizing radiation sources ❏ Lasers ❏ Welding ❏ Infrared generating equipment ❏ Ultraviolet generating equipment ❏ Sun	Radiation burns to: ❏ Eyes ❏ Body ❏ Skin	❏ Shaded safety glasses, with side shields ❏ Shaded safety goggles ❏ Welding helmet/goggles ❏ Protective clothing (welding leathers, gloves, etc.) ❏ Other
General safety hazards: ❏ Trip hazards ❏ Machine guarding ❏ Slip hazards ❏ Fall hazards ❏ Ladders ❏ Overhead hazards ❏ Broken equipment ❏ Electrical hazards ❏ Thermal equipment, process, or condition	Possible injuries: ❏ Foot and head injuries ❏ Lacerations, contusions, punctures ❏ Impact injury ❏ Crush injury ❏ Electrocution ❏ Thermal injury (extreme heat or cold) ❏ Amputation ❏ Sunburn	❏ Safety glasses with side shields ❏ Safety goggles ❏ Face shield ❏ Cut resistant gloves ❏ Hard hats ❏ Coveralls ❏ Work gloves ❏ Fall protection ❏ Thermal gloves ❏ Sunscreen ❏ Safety shoes
* Respiratory protection from TB and other airborne pathogens falls under 1910.134 and 1910.139 is therefore exempt from the hazard assessment requirements of this certification program.	** Operations that generate airborne fibers, mists, fumes, vapors, or dusts and require respirators or face filtering pieces are <u>not</u> required to be included in this assessment. A separate assessment and program, under 1910.134 and 1910.139, is required.	

OTHER (specify):

equipment that ensures a level of protection greater than the minimum required to protect the employees from the hazards. Take care to recognize the possibility of multiple and simultaneous exposure to a variety of hazards. Adequate protection against the highest level of each of the hazards involved in the task must be provided or recommended for purchase.

Protective Devices

Careful consideration should be given to comfort and fit of PPE in order to ensure that it will be used. PPE is generally available in a variety of sizes, and care should be taken to ensure the proper fit. Gloves that are too large, for example, do not provide adequate protection.

Overcoming employee objections to protective equipment may be one of the most challenging tasks. Many supervisors find that the objections they encounter are quite similar whether workers are talking about eye protection, respiratory equipment, or protective clothing, such as aprons or gloves.

One common complaint is that a particular piece of equipment is uncomfortable. No one wants to work with uncomfortable equipment or items that require readjustment every few minutes. Why is the eyewear or some other piece of equipment uncomfortable? Check to see if a qualified person properly fit the equipment.

Find out whether the employee had a choice in the selection process. Many employees are concerned about appearance as well as comfort. When possible, give employees a choice in the style or color of their PPE. Also, check to determine how easy it is for the employee to clean and maintain the protective equipment. This is particularly important in the case of respiratory protection.

There are many ways to minimize complaints and handle objections, such as learning as much as possible about the types of equipment available and networking with other medical practices about ideas for conforming to standards. In addition, safety equipment companies respond to today's changing business environment by providing PPE that incorporates rigid performance standards with good-looking colors, comfort, and style.

OSHA generally requires employers to provide and pay for necessary PPE. Most medical practices use disposable protective equipment, such as gloves, booties, head coverings, and fluid-resistant gowns and masks.

Employees must be required to wear PPE safely designed for the work to be performed and it must be maintained in a sanitary and reliable condition. When applicable, items must meet NIOSH recommendations (ie, general PPE guidelines,[1] respiratory protection[2]), American Society for Testing and Materials (ASTM) standards,[3] or the latest editions of the corresponding standards published by the American National Standards Institute (ANSI).[4]

Eye Protection

Vision loss can have a devastating effect on peoples' lives and work. According to preliminary data from NIOSH, Prevent Blindness America (a volunteer health organization) reports that each year from 650,000 to 700,000 eye injuries occur in American workplaces. Estimates of total direct and indirect costs of disabling eye injuries in the workplace each year are more than $1.25 billion.

Prevention of eye injuries requires that people who may be exposed to eye hazards wear protective eyewear. This includes employees, patients, visitors, contractors, or others who pass through an identified eye hazard area. Eye hazards include flying particles, liquid chemicals, chemical gases or vapors, potentially injurious light radiation, and body fluids.

Eye protectors must meet the following minimum requirements:

- Provide adequate protection against the particular hazards for which they are designed
- Be reasonably comfortable when worn under the designated conditions
- Fit snugly without interfering with the wearer's movements or vision
- Be durable
- Be capable of being disinfected
- Be easily cleaned
- Be kept in good repair

Contact lenses alone do not provide protection from on-the-job eye hazards. Employees who wear contact lenses should be extra cautious around gases, vapors, fumes, and dust that may get into or behind the lenses. Persons using contact lenses must also wear appropriate eye and face protection devices in a hazardous environment.

Prescription safety eyewear Employees who wear prescription lenses while working in eye hazard areas must either wear eye protection that incorporates the prescription into its design or find eye protection that can be worn over the prescription lenses without disturbing the proper position of the prescription lenses or the protective lenses. Only qualified optical personnel should fit prescription safety glasses.

Emergency eyewash The chances for full recovery from chemical damage to the eye decline rapidly after 15 seconds, making the placement of emergency eyewash stations critical. Eyewash equipment includes eyewash wall units, quick-pour bottles, and first-aid eyewash with eyecups. These provide a controlled flow of cool, clean water so workers can flush their eyes in case of irritation or injury.

In an emergency, employees may need to locate the nearest emergency eyewash without being able to see; therefore, the pathway from the workstation to the emergency eyewash station must be free of obstacles. It should also be in a well-lighted area identified with a sign.

Emergency eyewashes must meet the requirements of ANSI Z358.1.[5] The standard includes recommendations on positioning, water flow, nozzle activation, valves, and water temperature. Table 7.2 lists steps to meet emergency equipment needs.

Face Protection

Face shields provide additional protection against splashes. They are designed to protect the whole face and are often worn when there is any chance for exposure to acid and chemicals. Face shields are considered secondary eye protection and must be worn over safety glasses or goggles.

TABLE 7.2

Meeting Your Emergency Equipment Needs

Follow these steps to assure compliance with the applicable regulations:
1. Know the ANSI requirements.
2. Consult a professional to assist with your needs assessment.
3. Train employees in the proper use of equipment.
4. Maintain equipment and test its performance regularly.
5. Always enforce safety standards.

Reproduced with permission from *Occupational Health & Safety*, Emergency Showers & Eyewash, September 2000.

Eye protection with face shields may be required in medical practices where laser surgery is performed, in laboratories, or where medical equipment requires sterilization.

Fluid-resistant masks must be worn to protect mucous membranes of the nose and mouth during procedures and patient care activities that are likely to generate splashes or sprays of blood, body fluids, secretions, and excretions.

Hand Protection

The National Safety Council estimates that almost 20% of all disabling job accidents involve hand injuries. Common ways in which hands are injured include contact with harmful substances, sharp edges, or lasers and other devices that may cause burns. Skin contact is a potential source of exposure to toxic materials. Proper steps should be taken to prevent such contact. Gloves are available that protect workers from any of these individual hazards or any combination thereof.

Chemicals, such as solvents; mechanical forces, such as pressure; environmental factors, such as radiation; and biological irritants, such as bacteria can injure the skin on hands. Effects of contact injury include blisters, rashes, infections, and dry skin that can crack open and bleed. Prevention of contact injury includes washing hands frequently with soap and water, reading and following the substance warning labels, and knowing what hand protection to wear and when.

Employees should select gloves based on performance characteristics of the gloves, conditions, duration of use, and hazards that are present. One type of glove does not work in all situations. Examples of situations where gloves may be required include tasks where employees handle objects with sharp edges or have potential contact with chemicals or infectious substances. Gloves may also be used to prevent burns or provide shielding from radiation produced by lasers and other devices.

Latex or vinyl gloves should be frequently changed and inspected for punctures before wearing. Double-gloving is recommended to decrease the risk of exposure by penetration if it does not interfere with the task. Less permeable surgical latex gloves are recommended for use over polyvinyl gloves. Lead-lined gloves must be worn in the direct X-ray field.

The first consideration in the selection of gloves for use against chemicals is to determine the nature of the substances to be encountered. Read MSDSs and instructions and warnings on chemical container labels before working with any chemical. Recommended glove types are often listed in the MSDS section for PPE.

Nondisposable gloves should be periodically replaced, depending on frequency of use and permeability to the substance(s) handled. Gloves overtly contaminated should be rinsed and then carefully removed after use. For more detailed glove care, see Table 7.3.

In situations where gloves cannot be used, there are other ways to protect hands, such as water-repellent creams, solvent-repellent creams, sunscreens, hand pads, thumb guards, and so forth. When appropriate, industrial hygienists or other safety professionals should be consulted to help select alternatives.

Hand injury prevention The National Safety Council provides suggestions for protecting from hazards that can hurt hands. Reminders and suggestions include the following:

- Take off rings, watches, and bracelets before starting work.
- Use gloves that are job-rated for the task.
- Turn off the power when cleaning, inspecting, and repairing machines. Follow lockout/tagout procedures, including posting warning signs.
- Check tools and equipment for worn, broken, or dull cutting blades.
- Keep work areas clean. Do not use bare hands to sweep away items.
- Properly store tools so they do not fall.
- Keep hands where they cannot get crushed when carrying material through a doorway or using a handcart.
- Correctly use the right tool for the job.

The Bloodborne Pathogens Standard (Standard Number 29 CFR 1910.1030) requires health care workers to wear gloves to prevent exposure to blood or other potentially infectious materials (see chapter 3, Infection Control). Management must provide an appropriate alternative to latex gloves for employees who are allergic to commonly used latex

TABLE 7.3

A Glove Care Checklist

- Not all gloves offer the same protection. If you're not sure which glove to use, ask your supervisor.
- Check for holes at the tips and between fingers.
- Before you take off gloves, rinse them to wash away chemicals.
- Periodically clean and dry cotton work gloves. Make sure you clean them thoroughly, especially gloves with rough surfaces, to improve the grip.
- Keep a spare pair of gloves to use while the cleaned pair dries.
- Don't leave chemical gloves turned inside-out. This can trap chemical or vapors inside the gloves and rot the glove material.
- Don't store gloves with the cuff folded over. The crease weakens the glove material, and it can tear easily.
- Consult your supervisor before cleaning or washing chemical protective gloves.
- Replace damaged or worn gloves.

Reproduced with permission from National Safety Council, Hand Protection, 1990.

TABLE 7.4
Recommended Strategies—Risk Reduction for Natural Rubber Latex (NRL) Gloves

- Reduce unnecessary exposure to NRL proteins. Synthetic materials have been cleared for marketing as medical gloves by the FDA and can be used effectively for barrier protection against bloodborne pathogens.
- If selecting NRL gloves for worker use, designate NRL as a choice only in those situations requiring protection from infectious agents.
- When selecting NRL gloves, choose those that have lower protein content. Selecting powder-free gloves offers the additional benefit of reducing systemic allergic responses.
- Provide alternative suitable non-NRL gloves as choices for worker use (and as required by OSHA's Bloodborne Pathogens Standard for workers who are allergic to NRL gloves).
- Use powder-free gloves to reduce the dissemination of NRL proteins into the environment and decrease the likelihood of reactions by both the inhalation and dermal routes.
- When wearing hand protective equipment, including NRL gloves, avoid contact with other body areas, such as the eyes or face.
- Wash hands immediately after glove removal, which helps minimize powder and/or NRL remaining in contact with the skin.

Courtesy Occupational Safety and Health Administration, *OSHA Technical Information Bulletin: Potential for Allergy to Natural Rubber Latex Gloves and Other Natural Rubber Products,* April 12, 1999.

gloves. Table 7.4 outlines OSHA's recommended measures to reduce the risk of natural rubber latex allergy in workers. Further information about latex allergies is discussed later in this chapter.

Torso and Additional Body Protection
Impervious or low-permeability gowns should be worn to prevent contact with antineoplastic drugs, ribavirin, and blood or body fluids. Properly store contaminated gowns in the area of use until disposal and wash or discard of soiled gowns. Always wear lead-lined aprons in the X-ray field.

Lab coats, aprons, and leggings   Appropriate PPE must be worn while working in laboratories. Lab coats can be used to protect street clothing against biological or chemical spills as well as to provide some additional body protection. They should be removed before leaving the work area.

CDC/NIH guidelines recommend wearing protective coverings for biocontainment practices, such as wraparound gowns, scrub suits, or coveralls.[6] Plastic or rubber aprons and leggings should be worn when there is a potential for splashing. Determine the specific hazard(s) and the degree of protection required before selecting specialized PPE for laboratory personnel.

Head Protection
Head injuries are caused by falling or flying objects or by bumping the head against a fixed object. Head protectors, in the form of protective hats, must resist penetration and absorb the shock of a blow. The shell of the protective hat is hard enough to resist the blow and the headband and crown straps keep the shell away from the wearer's skull. Protective hats can also protect against electrical shock. Head protection, such as

hard hats, is not routinely needed in medical practices; however, employees exposed to overhead hazards and contractors engaged in construction at the medical practice may need hard hats. Hard hats must be worn at construction sites when hazards from falling or fixed objects or electrical shock are present. Under OSHA requirements, hard hats must meet specific ANSI standards.[7]

Foot Protection

There are many types and styles of protective footwear. Employees must wear protective footwear when working in areas where feet are exposed to electrical hazards, dangers due to falling and rolling objects, or objects piercing the sole. Rubber-soled shoes can act as insulation against electrical shock. Appropriate footwear with good traction should be worn for wet or slippery areas.

In delivery rooms and surgical areas, disposable shoe covers (ie, booties) must be available to protect against exposure to blood and body fluids. Periodic conductivity checks should be made on footwear and efforts should be taken to minimize the potential for static electricity in surgical areas.

Hearing Protection

Employees must be provided with hearing protection devices and directed to wear them whenever noise levels exceed limits allowed by OSHA.

Cleaning and Maintenance

All PPE must be inspected, kept clean, and properly maintained at regular intervals so that it provides the necessary protection. Cleaning is particularly important for eye and face protection where dirty or fogged lenses could impair vision. PPE should be distributed for individual use whenever possible and must not be shared among employees until it has been properly cleaned and sanitized.

Protective equipment wears out or becomes defective from causes such as misuse, poor maintenance, tears, and cracking. The passage of time and exposure to air are two other circumstances that can affect the performance and safety of certain types of protective equipment. Thoroughly check PPE that has been in the supply room or storage for several months.

Contaminated PPE that cannot be decontaminated must be disposed of in a manner that protects employees from exposure to hazards.

Employee Training

Employees required to wear PPE must receive training in the proper use and care of the equipment. Periodic retraining for both employees and supervisors must include the following:

- When PPE is needed.
- What equipment is needed.
- How to properly put on, take off, adjust, and wear the equipment.
- The limitations of the equipment.
- The proper care, maintenance, useful life, and disposal of the equipment.

After training and before performing work requiring the use of the equipment, employees must demonstrate that they understand the components of the PPE Program and how to properly use their required protective equipment or they must be retrained.

OSHA specifies that employers may rely on training provided prior to July 5, 1994, by a previous employer or the knowledge and ability an employee gained through prior experience.

Employers must maintain written certification that each affected employee has been trained and understands the PPE program. The certification must include the name of each trained employee, the date of the training, and the subject of the certification. If the employer relies on training that is provided by another employer or relies on the employee's prior experience, the certification must include the date that the employer determined that the prior training was adequate, rather than the date of the actual training.

Retraining

Employees must be retrained if there is reason to believe they do not have the understanding and skill required. Encourage a supportive atmosphere among employees so they feel comfortable contacting their supervisor if they have questions or need additional information to use their PPE. Retraining may also be required if changes in the workplace or in the types of PPE render previous training obsolete.

Recordkeeping

Written records must be kept of the names of persons trained, the type of training provided, and the dates when training occurred. The medical practice must maintain employee records for at least 3 years.

LATEX GLOVE ALLERGY

Because so much of what goes into making PPE involves synthetics, people with highly sensitive skin or allergies may develop disorders, such as latex allergies. Contact with latex is an occupational hazard for employees of medical practices because latex-containing medical devices abound in health care settings. Nearly 10 million individuals work in the health care field and an estimated 200,000 nurses have developed allergies.

People can develop allergies over time. Just because someone has never been allergic to a substance in the past, does not mean that he or she will never become allergic. Many factors may play a role in the development of allergies, from the toxins one breathes to an improper diet over a protracted period. As people grow older, the chemical composition of their bodies changes and what "was" may not be what "is."

Reactions to latex range from mild to fatal (though rare). Its uses are so widespread that it is nearly impossible to avoid contact. The following is a list of some products that contain latex:

- *Emergency Supplies:* catheters, disposable gloves, elastic bandages, protective sheets, stethoscopes, surgical masks, goggles, respirators, and intravenous tubing.
- *Office Supplies:* erasers and rubber bands.

- *Other Products:* tires, carpeting, swimming goggles, racket handles, and shoe soles.

Although suspected, latex sensitivity was little understood until the late 1970s. In the 1990s, it was recognized as an occupational "epidemic" for health care workers. The latex gloves nurses wore to protect themselves were causing illnesses, occasionally so severe that they were no longer able to work as nurses.

Even exposure to low levels of latex can trigger symptoms, often within minutes. According to NIOSH, mild reactions involve skin redness, hives, itching, and sneezing. More severe reactions include respiratory difficulties, such as runny nose, itchy eyes, scratchy throat, and asthma. In some reported cases, shock has been documented, but such a life-threatening reaction is seldom the first sign of latex allergy.

Employers must provide alternative PPE, such as hypoallergenic gloves, glove liners, or powderless gloves, to workers who are allergic to latex gloves. When using powder with some gloves, it can increase exposure and worsen the condition. Safety equipment companies respond to this requirement by regularly introducing new lines of gloves, such as vinyl and nitrile gloves, that prevent allergic reactions in sensitive individuals.

If latex gloves are unavoidable, it is recommended that oil-based hand creams or lotions be avoided unless they have been shown to reduce latex-related problems and maintain glove barrier protection. After removing latex gloves, wash hands with a mild soap and dry thoroughly.

According to NIOSH, the amount of exposure needed to sensitize employees to natural rubber latex is unknown, but when latex protein exposures are reduced, the rate of sensitization decreases.

High-risk employees should be periodically screened. NIOSH recommends that workers showing symptoms of latex allergy consult a doctor experienced in treating the problem, and workers with a known allergy should avoid latex exposures and wear a medic alert bracelet.

Individuals with minor reactions to latex gloves should try several different brands of gloves.

Latex has more than a dozen proteins in it, and the FDA does not know which of these actually initiates the allergic reaction. The FDA requires that certain labeling statements appear on medical devices, including device packaging containing natural rubber that may come into contact with humans. The labeling statements must alert users that a product contains either dry natural rubber or natural rubber latex. For products containing natural rubber latex, labels must state that the presence of this material may cause allergic reactions.

Infection Control Guidelines

The CDC provides recommendations to prevent latex sensitization and reactions among health care workers who use latex barriers for protection against transmission of infectious agents. According to CDC, avoiding latex-containing products is the primary means of preventing sensitization and reaction. Other recommendations include:

- Develop a protocol to evaluate and manage personnel with suspected or known latex allergy by establishing surveillance for latex reactions within the facility, purchasing gloves, and measuring the effect of preventive measures. Educational materials and activities should be provided on appropriate glove use and the manifestations and potential risk of latex allergy.

- Review information on the barrier effectiveness of gloves and consider worker acceptance (eg, comfort, fit) when selecting gloves for purchase.
- Maintain a list of all gloves used at the facility according to whether they do or do not contain latex.
- Evaluate personnel with symptoms suggestive of latex allergy. Use serologic tests only for workers suspected of having latex allergy.
- Avoid the use of all latex products by personnel with a history of systemic reactions to latex.
- Use nonlatex gloves for personnel with localized reactions to latex.
- Target interventions (eg, substitution of nonlatex gloves and powder-free gloves) to areas of the facility where personnel have acquired allergic reactions to latex.

CDC also advises that for personnel with symptoms of latex allergy, workplace restriction or reassignment may be necessary.

RESPIRATORY PROTECTION

In some medical practices, the PPE Hazard Assessment may identify harmful concentrations of airborne contaminants that exceed OSHA's permissible exposure limit (PEL) or NIOSH's recommended exposure limit (REL) (see chapter 4). When those situations occur, steps must be taken to protect the health of employees. In most cases engineering controls (eg, ventilation, isolation, product substitution) can be used to reduce those concentrations and eliminate the hazard. However, in some cases it may be necessary to provide employees with a respiratory protective device (ie, respirator).

Do not substitute respirators for proper control measures. Only use respirators when effective engineering controls are not technically feasible, while controls are being installed or repaired, or when emergency and other temporary situations arise. In essence, OSHA has said that respirators are to be used as a last line of defense to protect the worker, and only when other control methods have been evaluated and found ineffective. The use of respirators is not to be taken lightly. Potential health effects associated with wearing respirators include poor regulation of body temperature, diminished senses (ie, decreased hearing, decreased voice clarity, reduced visual fields), psychological effects to the discomfort of wearing a respirator, and allergic skin reactions.[8]

Respirators are the least preferred method of worker protection from respiratory hazards because they can be unreliable if the employer does not establish an adequate respiratory protection program and if the employee does not cooperate. Respirators are not natural to wear and are often an imposition upon the wearer. Respirators can cause stress and discomfort and be a nuisance at times. Wearers should be reminded that respirators are being used because of a very real threat to their health.

Employees often need encouragement and motivation to continue using respirators safely and effectively.

Many variables may influence the selection of the appropriate exposure limit for a given contaminant, such as the specific situation, the employee, or the job. For example, the effects of some hazardous substances may increase due to exposure to other contaminants present in the workplace or the general environment or to medications or personal habits of the employee. Such factors, which affect the toxicity of a contaminant, would not have been considered in the determination of the specific exposure limit. Also, some substances are absorbed by direct contact with the skin and mucous membranes, thus potentially increasing the total exposure.

Effective respirators must be used as outlined in OSHA's Respiratory Protection Standard (Standard Number 29 CFR 1910.134). If respirators are used in the medical practice, employers must develop and implement a written respiratory protection program with required worksite-specific procedures and elements for respirator use. The program must be administered by a trained program administrator and periodically evaluated and updated as necessary. Written program requirements are provided in OSHA's standard and include the following:

- Procedures for selecting respirators.
- Medical evaluations of employees who are required to use respirators.
- Fit testing procedures for tight-fitting respirators.
- Procedures for proper use of respirators in routine and reasonably foreseeable emergency situations.
- Procedures and schedules for cleaning, disinfecting, storing, inspecting, repairing, discarding, and otherwise maintaining respirators.
- Procedures to ensure adequate air quality, quantity, and flow of breathing air for atmosphere-supplying respirators.
- Training of employees in the respiratory hazards to which they are potentially exposed during routine and emergency situations.
- Training of employees in the proper use of respirators, including putting on and removing them, any limitations on their use, and their maintenance.
- Procedures for regularly evaluating the effectiveness of the program.

Respiratory Protection for M. tuberculosis

For medical office staff required to wear respirators for protection against tuberculosis, the OSHA requirements of Standard Number 29 CFR 1910.134 apply until OSHA's proposed tuberculosis standard is finalized into law. The OSHA Regulation (Standard Number 29 CFR 1910.139) can be used to identify the type of face filtering respirator (ie, such as the N95, N99, or N100) the medical practice will use.

ENDNOTES

1. National Institute for Occupational Safety and Health (1997). *Personal Protective Equipment Program*. CDC Program. Available at: www.cdc.gov/niosh/. Last modified January 2, 1997.

2. National Institute for Occupational Safety and Health (1999). *Respiratory Protection Program in Health Care Facilities, Administrator's Guide*. US Department of Health and Human Services, Public Health Service, Centers for Disease Control and Prevention, National Institute for Occupational Safety and Health; Cincinnati, OH.
3. American Society for Testing and Materials (ASTM); Philadelphia, PA.
4. American National Standards Institute; New York, NY.
5. American National Standards Institute. ANSI Z358.1. *Emergency Eyewash and Shower Equipment*.
6. Centers for Disease Control and Prevention. *Personal Protective Equipment Program*. Available at: www.cdc.gov. Last Modified: 1/2/97.
7. American National Standards Institute. ANSI Z 89.1-1986, *Safety Requirements for Industrial Head Protection*, (if purchased after July 5, 1994) or ANSI Z89.1-1969 (if purchased before July 5, 1994), New York, NY.
8. National Institute for Occupational Safety and Health (1987). *NIOSH Respirator Decision Logic*. DHHS (NIOSH) Publication No. 87-108.

chapter 8

Hazardous and Medical Waste Management

Before the late Rachel Carson's Silent Spring *was published in 1962 decrying what humankind was doing to the environment (and long before the EPA was established in 1970), the average American paid little attention to its trash or garbage. The general attitude was that this world had an inexhaustible supply of natural, "self-renewable" resources. Nature would heal itself. That blithely naïve attitude has since received so many warnings to the contrary that it is no longer possible to avoid the consequences of unthinking, careless behavior.*

An ever-growing body of requirements for handling hazardous and medical wastes regulates medical practices. Numerous federal, state, and local rules apply to the management of materials that may pose risks to human health or the environment. In addition, voluntary industry compliance bodies prepare standards to provide guidance.

Involved health care personnel may be aware of the existence of such responsibility; however, materials and waste dangers and the sources of the regulations that apply can be confusing.

Two goals that are of primary importance in the proper handling of hazardous and medical wastes are that compliance must be achieved and the program must be made cost-effective.

CATEGORIES OF DANGEROUS HEALTH CARE WASTE

There are a number of terms used to categorize wastes such as hazardous, infectious, biological, chemical, radioactive, and chemotherapeutic, but they are often confusing. Generally, there are three categories of dangerous health care materials:

1. *Hazardous Chemical Wastes.* These include the various wastes produced by hazardous chemicals used in the medical practice, such as glutaraldehyde, phenol, mercury, and cytotoxic chemicals. Hazardous materials are not considered wastes until they are thrown away.
2. *Medical (Infectious) Wastes.* Medical waste is defined as waste that has the potential to transmit infection and includes a number of wastes generated in patient care and treatment, such as used and unused sharps, bulk human blood and blood products including items saturated with blood, and pathological wastes.

3. *Radioactive Wastes.* This is a separate category of waste that must be integrated into the overall medical waste management program and requires separate procedures. It includes radioisotopes used in diagnosis and treatment of patients. Clinical laboratories and radiology departments use radioactive materials. This category of waste is not covered in this document. Standards and guidance from regulatory agencies, such as the state department of health or the NRC,[1] should be consulted.

REQUIREMENTS FOR ENVIRONMENTAL MANAGEMENT

Throughout the last two decades, growing concern has centered on the practices of business and industry that jeopardize the environment and human health. During that time, a number of agencies have been created within government to promulgate rules and develop compliance guidelines. Several sets of guidelines apply. Agencies and organizations involved in the environmental-responsibility movement include EPA, OSHA, CDC, DOT, and JCAHO. Chapter 2, The Creation of Occupational Safety and Health Agencies, summarizes these agencies and organizations.

THE PROCESS OF WASTE MANAGEMENT

While the myriad regulations may appear confusing and disjointed, taken together they can be seen as complementary rules that form a nexus of control and protection for workers, the community, and the environment. Some health care professionals may believe that certain elements of this collection of regulations are unimportant or irrelevant to their situation; however, collectively, they guide the medical practice in a well-reasoned process of hazardous materials management.

For instance, OSHA's Hazard Communication Standard (commonly known as the "Right-to-Know Law" or HAZCOM), which is discussed in chapter 4, Chemicals, launches the ongoing process of tracking chemicals as they enter and disperse throughout the facility. It helps to protect employees by providing them with important information while they use and work around the hazardous chemicals.

At the same time, state medical waste guidelines and rules require specific handling, disposal, and safety procedures. The result is that medical waste management is integrated into a dynamic step-by-step process. The total plan of control is composed of several elements that feed off each other. Viewed from this perspective, the hazardous materials and medical waste standards become a manageable collection of rules that assist in organizing and integrating all waste management activities. This in turn protects the medical practice from costly liability and fines for noncompliance.

LIABILITY ISSUES

Noncompliance with mandatory regulations is a serious matter and can carry fines and civil and criminal penalties and can place employees and patients at risk. The agencies with authority can make warnings, exact fines, initiate civil lawsuits, or begin criminal proceedings against responsible personnel in facilities that do not comply.

A growing area of concern for health care facilities is "environmental impairment liability." This term refers to the legal recourse given in many cases to members of the community and governmental personnel to recover damages for harm to the environment or human health caused by the health care provider. This includes liability for the waste and other material *after* it leaves the facility. This potential liability is why it is so important for medical practices to be certain not only that they comply, but also that their transportation and disposal companies comply and are vigilant in their handling and disposal practices.

HAZARDOUS WASTE

When thrown away, hazardous chemicals become hazardous wastes because the materials pose risks to the medical practice and environment. They must be appropriately contained and stored. Improper disposal can pollute the environment and endanger the life and health of people and wildlife in the community. Mismanagement of hazardous wastes, such as pouring mercury down the drain or discarding glutaraldehyde solutions into a septic system, can create immense complications for medical practices.

EPA is given the power under several laws to prevent such harm to the environment and exercises control over the release of harmful materials into the environment. Its rules and recommendations define what substances can be hazardous to human health and the environment. The law of primary interest to medical practices is the Resource Conservation and Recovery Act of 1976 (RCRA). Its mission is to protect human health and the environment. RCRA regulates the management of hazardous waste using a "cradle-to-grave" approach. RCRA directs EPA to oversee a national program to control hazardous waste from the time it is generated to its final disposal. Federal hazardous waste regulations are located in Title 40 of the Code of Federal Regulations (CFR) Parts 260 to 262[2]. In addition, EPA provides numerous reference materials including, *Understanding the Hazardous Waste Rules: A Handbook for Small Businesses.*[2]

While most medical practices generate small amounts of hazardous wastes and are covered by EPA's minimum requirements, the following should be included in a hazardous waste management program to minimize hazards, reduce liability, and comply with regulations:

- Identify hazardous waste
- Survey for hazardous waste
- Determine generator status
- Segregate waste
- Store waste
- Label waste
- Dispose of waste
- Plan for emergencies
- Reduce waste

Identify Hazardous Waste

Hazardous wastes carry serious threats to humans and the environment. A wide variety of health problems have been associated with some wastes. Others can explode, ignite, or harm the environment. Along with these threats comes the possibility for liability and regulatory noncompliance. A major step toward limiting these sources of risk is identifying the hazardous wastes present in the facility. RCRA sets forth criteria for identifying waste as hazardous. Table 8.1 provides the RCRA criteria.

A first look at these definitions may cause alarm. How can anyone know if a chemical will ignite or react? But if a total waste management system is in place in the medical practice, deciding which wastes fit which category is relatively easy. The material safety data sheet (MSDS) for each hazardous chemical (which comes with the chemical from the manufacturer) lists the potential for fire, reactivity, and other possibilities. In addition, the HAZCOM provides requirements to ensure that chemicals are properly labeled.

Survey for Hazardous Waste

Once identification criteria are understood, the first priority is to conduct a survey to determine if hazardous wastes are present and, if so, into what category do they fall. Waste transporters, handbooks from government agencies, and private organizations can be helpful.

Using the survey as well as information from other sources, data concerning a hazardous waste can be properly recorded. A form should be developed for this purpose that lists the department, name and title of person reporting the information, and date. The form should ask for the name of the chemical, manufacturer, average amount stored, quantity of chemical generated as waste (ie, per day, week, or month), the concentration, and the current method of disposal.

Determine Generator Status

After completing the survey, synthesize the results into a portrait of the waste-generating and disposal practices of the medical practice. From this, determine the medical practice's generator status (ie, classes, such as corrosive or toxic, and amount, such as how many pounds).

TABLE 8.1

Classifications of Hazardous Wastes

Classes	Descriptions	Examples
Corrosive	A chemical capable of dissolving or wearing away of substances, including body tissue.	Glutaraldehyde
Flammable	Easily catches fire and tends to burn rapidly.	Phenol, formaldehyde in its vapor state
Reactive	A reactive chemical is capable of participating in a rapid and sometimes violent reaction with other chemicals and substances. For example, mixing formaldehyde and phenol.	Formaldehyde
Toxic	Can cause injury or death if swallowed, inhaled, or absorbed through the skin.	Antineoplastic drugs, mercury, glutaraldehyde, phenol

Courtesy the Environmental Protection Agency, *Understanding the Hazardous Waste Rules: A Handbook for Small Businesses*, 1996 (update).

EPA recognizes three categories of generators, as listed in Table 8.2. Each of the three categories has specific EPA regulations the medical practice must understand. Many medical practices are affected by RCRA because they generate hazardous waste; however, they may be classified as conditionally exempt generators and have less burdensome requirements.

Segregate Waste

After determining the medical practice's generator status, specific procedures should be developed for complying with regulations and maintaining a safe work environment. One procedure is segregation, or source separation.

If nonhazardous material is not separated from hazardous waste before it is stored to be discarded, then it becomes hazardous waste and should be marked as such. This can greatly increase the amount of waste that is classified as hazardous, consequently multiplying the cost of treatment, transportation, and/or disposal. Therefore, it is important that hazardous waste be separated from other wastes from the start.

Store Waste

Due caution must be exercised in storing hazardous wastes. These waste materials cannot be stored without a RCRA permit unless meeting one of the following criteria:

- For large-quantity generators, they are kept no longer than 90 days.
- For small-quantity generators, they are stored for no longer than 180 days, or 270 days if the waste is to be shipped more than 200 miles.
- The medical practice is classified as a conditionally exempt generator.

RCRA strictly regulates storage containers and conditions. The regulations include inspections, security, maintaining appropriate room conditions, and accurate labeling. Examples of commonsense storage requirements include maintaining containers in good condition, keeping containers closed except when filling or emptying them, and conducting weekly inspections of the containers for leaks or corrosion. Never store

TABLE 8.2

Categories of Generators

Conditionally Exempt Generator (A classification of Small Quantity Generator)

Up to 100 kg (about 220 lbs. or 25 gallons) of nonacutely hazardous wastes per month and 1 kg (about 2 lbs.) of acutely hazardous waste in any calendar month.

Small Quantity Generator

Between 100 kg (220 lbs.) and 1,000 kg (2,200 lbs.) of nonacutely hazardous wastes per month.

Large Quantity Generator

Greater than 1 kg of acutely hazardous wastes per month and greater than 1,000 kg of nonacutely hazardous wastes per month.

Reproduced with permission from *Managing Health Care Hazards, Compliance Publication Series*, published by Chaff & Co. Copyright 1993.

different wastes in the same container that could react together to cause fires, leaks, or other releases.

Compatibility

Incompatible hazardous wastes, when stored together, can cause dangerous chemical reactions, such as igniting or exploding. For this reason, RCRA requires that items of different hazard classes be packed into separate containers. Reference sources, such as MSDSs, are available that can help determine waste characteristics and how different waste classes should be stored. Table 8.3 lists general chemical storage compatibility.

Even after they have been separated at the source and packaged, hazardous wastes may be incompatible and cause dangerous reactions in close storage depending on their characteristics (eg, ignitability, corrosivity, reactivity, toxicity).

Label Waste

If a medical practice is classified as a conditionally exempt generator and is storing hazardous waste, waste containers must be clearly labeled with the words "HAZARDOUS WASTE" and list the contents and date collection began in the container. Standard EPA labels that contain additional information are available from many suppliers and label companies.

Dispose of Waste

The person in charge of hazardous waste disposal should be involved enough in the purchasing process to ask about the disposal process.

The medical practice must have policies and procedures for preparing its waste for disposal. After establishing these steps, the major issue

TABLE 8.3

Chemical Compatibility and Segregation Information

General Chemical Storage Compatibility						
+ means these groups may be stored together in most cases				* means store these groups AWAY from water and water sources		
Group 1 +	Group 2 +	Group 3 +	Group 4*	Group 5*	Group 6*	Group 7*
Hologenated compounds	Ketones	Organic acids	Amines and alkanolamines	Caustics	Oxidizers	Inorganic acids
Olefins	Saturated hydrocarbons	Acid amhydrides	Ammonia	Hydroxides	Nitrates	Hydrochloric acids
Alcohols, glycols, and glycol ethers	Aromatic hydrocarbons	Acetic acid		Carbonates	Persulfates	Sulfuric acids
Phenol	Oils					Phosphoric acids
Chloroform	Aldehydes					Halogens
Dyes and stains	Olefins					
Ethidium bromide	Esters					
	Formaldehyde					

Courtesy the National Institute of Health, Chemical Compatibility and Segregation Information.

becomes, "What is the best method of disposal?" If the medical practice is unsure of acceptable disposal practices, consult the local wastewater or sewage treatment office and the state hazardous waste management agency to check on disposal preferences and regulations. The following are some of these methods:

- In some situations, the medical practice stores the hazardous waste to be shipped off-site for treatment or disposal. This method is the most frequently used option.
- Other disposal alternatives include recycling or on-site treatment and disposal.
- Distillation, reclamation, and neutralization are also options, but these options are not usually used in medical practices due to equipment costs and elaborate permitting requirements.
- Small quantities of some hazardous waste solutions may be discharged directly to a publicly owned treatment works (POTW) without being stored or accumulated first. This discharge to a POTW must comply with the Clean Water Act. POTWs are public utilities usually owned by the city, county, or state that treat industrial and domestic sewage for disposal. If using this process, pour the solutions down the drain along with cold running water (for approximately 5 minutes) to dilute the solution's concentration. However, this is often not considered good management practice, even if allowed, and in many communities, it may be illegal. Consult local regulatory agencies regarding this practice.
- Hazardous waste should never be discarded into a septic system.

Cytotoxic Waste

The toxicity of antineoplastic agents is inherent because of its effectiveness against cancer. Because individual responses to drugs in this category widely vary, many different antineoplastics have been marketed. Exposure of health care personnel must be minimized. At the same time, the requirement for maintenance of aseptic conditions must be satisfied.

Chemotherapy and antineoplastic chemicals may account for the largest volume of hazardous wastes produced by medical practices. The greatest volume of antineoplastic wastes is generated from drug-dispensing devices, contaminated protective clothing, and associated paraphernalia, such as needles.

Cytotoxic waste disposal requires specific measures, including storage and labeling. Disposal of cytotoxic drugs and trace-contaminated materials (eg, gloves, gowns, needles, syringes, vials) presents a possible source of exposure to nurses and physicians as well as to ancillary personnel, especially the housekeeping staff. Excreta from patients receiving cytotoxic drug therapy may contain high concentrations of the drug. Employees should be aware of this source of potential exposure and take appropriate precautions as established by the medical practice to avoid accidental contact. OSHA guidance[3] specifies health and safety guidelines for personnel dealing with cytotoxic (ie, antineoplastic) drugs. The guidelines include recommendations regarding work areas, prevention of employee exposure, medical surveillance, hazard communication, training and information dissemination, and recordkeeping.

Plan for Emergencies

A contingency plan is a plan that attempts to look ahead and prepare for any accidents that could possibly occur. It can be thought of as a set of answers to a series of "what if" questions. For example: "What if there is a spill of hazardous waste or one of the containers leaks?" Emergency procedures are the steps to follow if there is an emergency. These procedures are also helpful for informing employees about their responsibilities in the event of an emergency.

Emergency phone numbers and locations of emergency equipment should be posted near telephones, and employees must know proper waste handling and emergency procedures. An employee should be appointed to act as emergency coordinator to ensure that emergency procedures are carried out in the event an emergency arises.

Reduce Waste

There is an important and effective waste management strategy often overlooked—waste reduction. An ongoing program can reduce the amount of hazardous waste the facility generates. Some processes can be avoided or altered to result in less waste. The strategy to reduce waste should include the following:

- Gathering information about current practices concerning generation, handling, storage, and treatment.
- Disposing of waste, after carefully evaluating existing waste-reduction practices.
- Developing recommendations for waste reduction through source control, treatment, and recycling techniques.
- Assessing costs and benefits of existing and recommended waste-reduction techniques.

Another alternative is substitution. Increasingly, manufacturers are producing less hazardous substitutes for products that previously contained hazardous chemicals. After identifying hazardous chemicals used in the medical practice, contact the manufacturers or distributors of the chemicals or chemical compounds to determine if they have nonhazardous or less hazardous equivalents. The number of possible substitutions may be surprising. Even if they are more expensive, the cost is probably made up in savings on waste storage, treatment, and disposal.

A third way to reduce the amount of waste requiring special disposal procedures is neutralization, or "denaturing." This process converts hazardous substances to relatively harmless ones that can be disposed of in a routine manner such as through the sewer system. Only trained personnel and those familiar with the RCRA Standard should perform such tasks.

Other ways to reduce the amount of generated hazardous waste include the following:

- Do not mix nonhazardous wastes with hazardous ones. For example, do not put nonhazardous cleaning agents or rags in a container of hazardous waste or the entire contents become subject to the hazardous waste regulations.

- Avoid mixing several different hazardous wastes. Doing so may make recycling very difficult, if not impossible, or make disposal more expensive.
- Avoid spills or leaks of hazardous products. The materials used to clean up such spills or leaks also become hazardous.
- Make sure the original containers of hazardous products are completely empty before thrown away. Use the entire product.
- Establish an internal recycling program.
- Avoid using more of a hazardous product than is needed. For example, use no more glutaraldehyde than is needed to do the job.
- Purchase small quantities of new hazardous chemicals (eg, cleaning compounds) until it is determined that they perform as expected and do not need disposal.
- Encourage drug and chemical suppliers to become responsible partners in a waste-minimization program by ordering from suppliers who provide quick delivery of small orders, accept return of unopened stock, and offer off-site waste management outlets or cooperatives for hazardous wastes.

These practices apply to all waste streams. Reducing hazardous waste means saving money on raw materials and reducing the costs to the medical practice for managing and disposing of them. The following sections provide additional waste-minimization methods for certain waste streams in medical practices.

Reducing Cytotoxic Wastes

There is significant potential for reducing cytotoxic waste volumes through administrative controls. These include waste segregation, minimizing cleanup waste volume, and employee training. Other methods include the following:

- Segregate chemotherapy wastes from other wastes. Provide separate containers with distinctive labels in chemotherapy drug-handling areas.
- Discard disposable garments with nonhazardous refuse if no chemotherapy agents were spilled during handling. However, assume gloves are contaminated.
- Minimize the cleaning frequency and volume of gauze material used for the compounding hood.
- Purchase drug volumes according to need. Over-purchasing results in the generation of out-of-date materials that must be discarded. Reducing the generation of residual material may be accomplished by computing daily compounding requirements of each drug and ordering appropriately sized containers. In addition, obtain prescored ampoule containers. This minimizes spillage associated with breaking open unscored ampoule necks.
- Employ proper spill containment and clean-up procedures. Spill containment and clean-up kits should be readily available in the compounding area(s). These kits, usually available from the drug suppliers, should contain both small and large absorbent devices.

- Implement proper training. Effective administrative and engineering controls require ongoing employee training and supervision. In addition to safety and health training regarding chemotherapy drug handling, employees should also be trained in methods to minimize generation of chemotherapy wastes.
- Return outdated drugs to the manufacturer.
- Centralize the location of chemotherapy compounding areas.

Reducing Formaldehyde Wastes

Ways to minimize formaldehyde wastes include installation of reverse-osmosis water supply equipment, using minimum effective cleaning procedures, recycling and reusing waste solutions, and proper waste management.

All waste management methods should stress control of airborne emissions because formaldehyde is a suspected carcinogen of the upper respiratory system.

Reducing Mercury Wastes

Perhaps the best, if not the least costly, approach to mercury waste minimization is to eliminate mercury-containing instruments entirely. Many medical practices substitute solid-state electronic sensing devices for mercury-based thermometers and blood pressure instruments. This source elimination technique appears to be the primary reduction alternative for mercury wastes. The higher initial costs of electronic devices are typically justified because they eliminate costly clean-ups and associated hazards from glass breakage and mercury spills.

Elemental mercury exhibits high toxicity via inhalation, skin absorption, and ingestion. Spill clean-up procedures and handling operations must be carefully designed and monitored to protect employees and public health. Specially designed mercury vacuums and spill-absorbent kits should be used for spill clean-ups.

Waste mercury can easily be recycled depending on the type or degree of contamination. Proper safety controls must be utilized. Check with the regional EPA office or state environmental department for information about the area's commercial mercury-recovery firms. Mercury refineries in the United States are also listed in resource documents and on the Internet. Some mercury recyclers provide containers for collecting and shipping.

REGULATED MEDICAL WASTE

Public concern about medical and infectious wastes heightened after vials of blood and syringes washed up on East Coast beaches during the summers of 1987 and 1988. As public concern grew, a number of actions led by government private groups developed to address waste management. Many states acted quickly to pass legislation governing various aspects of medical and infectious waste management, creating a variety of regulatory programs that vary widely from state to state.

In the early 1990s, approximately 158 million tons of medical waste, representing 0.3% of total waste, was generated annually in the US. Medical waste may be generated by numerous sources, including clinics (eg, chronic dialysis, free clinics, community, surgical, urgent care, abortion, drug rehabilitation, health maintenance organizations), all specialties of medical practices, emergency medical services, medical and nursing schools, and health units in industry, schools, and correctional facilities.

The Medical Waste Tracking Act of 1988 (MWTA) required the EPA to identify alternative (ie, nonregulatory) approaches to medical waste management.

EPA divided medical waste into two broad categories: waste that has the potential to transmit infection and waste that poses a risk to public health or the environment for reasons other than infectious potential. EPA defined medical waste as any solid waste used in health care facilities that, if not contained and managed properly, *could* result in the transfer of infection. It did not include wastes regulated as hazardous under RCRA, Subtitle C—*Managing Hazardous Waste*.[4] These definitions could include any or all of the following medical practice-generated waste items:

- *Microbiological wastes*: cultures and stocks of infectious agents and associated biologicals
- *Blood and blood products*: serum, plasma, other blood components
- *Pathological wastes*: tissues, organs, and body parts removed during surgery or autopsy
- *Used sharps*: hypodermic needles, syringes, Pasteur pipettes, broken glass, scalpel blades
- *Animal wastes*: contaminated animal carcasses; body parts; bedding of animals exposed to infectious agents during research, production of biologicals, or testing of pharmaceuticals
- *Communicable disease, CDC Class 4, isolation wastes*: biological waste and discarded materials contaminated with blood, excretion, exudates, or secretion from human beings or animals who are isolated to protect others from communicable diseases
- *Contaminated lab wastes*: slides and cover slips, disposable gloves, laboratory coats, aprons
- *Surgery and autopsy wastes that were in contact with infectious agents*: soiled dressings, sponges, drapes, lavage tubes, drainage sets, underpads, surgical gloves
- *Dialysis wastes*: free-flowing blood or blood components; used sharps; filter units; disposable sheets, towels, gloves, aprons, and laboratory coats
- *Contaminated medical equipment* (if discarded and was in contact with infectious agents)
- *Unused discarded sharps*

Some confusion existed regarding EPA's definition of medical waste and CDC's definition of infectious waste. CDC defines infectious waste as microbiological waste (eg, cultures stocks), blood and blood products, pathological waste, and sharps. Many state regulatory agencies and health care associations have come to agreement on a reasonable definition for infectious waste, which closely parallels the CDC definition as follows:

- Microbiological wastes
- Blood and blood products
- Pathological wastes
- Used sharps
- Class 4 communicable disease isolation wastes
- Contaminated animal carcasses

Regulatory bodies claim that the other categories mentioned under MWTA are covered under one or more of these six categories.

Later terminology uses the word "medical," reserving "infectious" for waste known to be capable of transmitting disease. In this chapter, except for direct quotes, the term "medical waste" is used throughout to refer to wastes that are potentially infectious or that pose a potential threat to the public health and safety.

Medical waste disposal is now regulated by each state with widely varying requirements. Most states require records to account for the amount of medical waste that was generated. This and other requirements are spelled out in the waste permit. Read the permit carefully to avoid compliance problems. These records provide liability protection to the generator as well as documentation of good waste management practice.

A big concern for a medical practice is the possibility that improperly disposed of regulated medical waste may bring about fines or public relations problems. Ways to prevent these problems from occurring include effective recordkeeping and choosing qualified waste transporter and disposal firms.

Planning the Medical Waste Program

In the rush to begin a medical waste program, discovering existing risks and conditions can often be a neglected step in the process. To alleviate this oversight, medical practices must carefully plan their medical waste programs. Enthusiastic support for the program can enhance work ethics among the staff for handling the waste and can also provide a public relations boost for the medical practice when the waste culture changes.

The first step is to get a clear picture of the existing practices in the facility. Identify risks to determine medical waste management needs. Interact with employees, and encourage them to suggest their ideas for the program.

Council of State Governments Medical Waste Guidelines

Proper medical waste handling begins with proper management of generation sites. Design of waste containers and storage areas affect the quality of handling. In addition, employees should be properly trained. The CSG entered into a grant agreement with the EPA's Office of Solid Waste in 1990-1991 to develop guidelines for use by states and other entities that generate and/or manage medical waste. The following guidelines are a result of that effort and can be found in CSG's document, *Model Guidelines for State Medical Waste Management*[5]:

- Segregation
- Handling

- Containment
- Labeling
- Storage (Prior to Treatment)
- Monitoring and Recordkeeping
- Training
- Contingency Planning
- Transportation

Segregation

Segregation is the initial and crucial point in the waste-handling process that determines the amount of waste and type of treatment to which the waste is subjected. Important steps in the segregation process include:

- Designate medical wastes as soon as practical at the point and time of origin.
- Separate medical waste from other solid waste, such as paper and garbage items.
- Separate sharps.
- Separate other medical wastes destined for on-site treatment from those destined for off-site treatment.
- Separate wastes intended for recycling.
- Segregate according to treatment method and packaging suitable for that method:
  - liquid
  - sharps
  - nonsharps and other solids, according to heat value, moisture content, and biological and chemical composition
  - Provisions should be made for separating medical waste with multiple hazards (eg, radioactive sharps) when additional or alternative treatment is required.

Handling

The following handling techniques can assure the safety and protection of employees from injury:

- Compact untreated medical wastes only if the compaction takes place in a closed chamber that eliminates the possibility of exposure to infectious agents through aerosols.
- Do not transfer medical wastes through chutes or dumbwaiters, as these practices may force contaminated air into other areas of the building such as nursing stations or other clinical areas. Such handling also may lead to breakage of containers in the waste area and contamination of the waste shaft and openings with potentially infectious material.
- Frequently disinfect carts used to transfer wastes within medical practices and do not use them for other materials.

Containment

Proper containment of waste during storage, transport, treatment, and disposal can reduce the chances of exposure to infection. In addition, proper containment of medical waste protects employees from physical injury, reduces the possibility of unaesthetic appearances, and speeds up the waste handling process. Recommended container practices include the following:

- Clearly marked, easily accessible containers for each type of waste encourages optimal segregation.
- Locate the containers in the immediate area of use.
- Containers can include recyclables and reusables if the medical practice's policy on infectious control and liability concerns allows for it.
- Too many containers can confuse and discourage personnel from attempting to properly designate the various wastes. Too few containers results in all wastes being designated for the costly and more involved process necessary for only certain types of waste.
- Routinely replace the containers and do not allow them to become overfilled.
- Seal all bags by lapping the gathered open end and binding it with tape or a closing device so that no liquid can leak.
- If the outside of the bag is contaminated with body fluids, use a second outer bag. Use double plastic bags with liquid wastes.
- Thoroughly wash and decontaminate containers that will be reused for medical waste. Employ an approved method each time they are emptied unless the surface of the containers are protected with a disposable liner or other means. Scrub reused containers to remove any visible solid residue, then disinfect.
- Enclose and store incinerator ash in tightly lidded containers or sturdy plastic bags so that employees are not exposed to inhalable dust or spill when transferring the ash.
- Select packaging materials appropriate for the type of waste and treatment process. Use packaging that maintains its integrity during storage and transport.
- Do not use glass containers as primary containers for transportation of medical waste. Place glass containers into a rigid or semi-rigid, leakproof container to protect from breakage.
- Use plastic bags that are impervious to moisture, puncture resistant, and distinctive in color or markings.
- Reusable containers should be constructed of either heavy-wall plastic or noncorrosive metal. Do not use these containers for any other purpose unless they have been properly disinfected and have had medical waste symbols and labels removed.
- Support heavy materials in double-walled corrugated fiberboard boxes or equivalent rigid containers.
- Place liquid/pourable wastes in leak-proof, rigid, puncture-resistant, break-resistant containers that are capped or tightly stoppered bottles or flasks.
- Place needles, syringes, breakable items, and other sharps in a plastic vial or puncture-resistant box before placing them into the bag. Do not clip or recap needles by hand. Do not crush syringes.

Labeling

Clearly and immediately mark each container to be transported off-site as medical waste after packaging.

- The label or tag should also identify the generator, transporter, and date of shipment.
- The legible label should be securely attached to the outermost container with string, wire, adhesive, or other method that prevents loss or unintentional removal.
- Use indelible ink to complete the information on the label, which should measure at least 3 inches by 5 inches in size. The lettering for "medical waste" should be no less than an inch in height. The wording should be readily visible from any lateral direction when the container is upright.
- Include the following information:
  - The name, address, business telephone, and state permit or identification number (if applicable) of the generator.
  - "Biomedical Waste" or "Medical Waste" in large print.
  - The name, address, business telephone, and state permit or identification number (if applicable) of all transporters, treatment facilities, or other persons to whose control the medical waste is transferred. License number of transporters must be provided if applicable.
  - The international biohazard symbol should appear on every container. (See Figure 8.1.)
  - The date upon which the medical waste was packaged.
- Label treated medical waste with the following information:
  - The name, address, and business telephone number of the generator.
  - The date upon which the medical waste was treated.
  - Treatment method utilized.
  - Statement indicating that the waste has been treated and is no longer medical waste.

FIGURE 8.1

The International Biohazard Symbol

Storage (Prior to Treatment)

Medical wastes may need to be stored on-site until a large enough quantity is accumulated to warrant treatment at the facility or until collection for transport to an off-site treatment facility is scheduled. Treatment system malfunction or staff shortages may also necessitate storage of waste. In addition, intermediate storage facilities may be necessary en route to off-site facilities. Rural areas may find that the use of transfer or collection stations expedites proper management by small generators. All storage areas or units should be well secured to discourage the theft of needles by drug users and to keep animals away from organic matter (ie, body parts) that is contained in the waste.

The following conditions apply to storage, transfer, and collection stations:

- Store medical waste in a specifically designated area located at or near the treatment site, or at the pickup point if it must be transported off-site for treatment.
- All areas used to store medical waste should be durable, easily cleanable, impermeable to liquids, and protected from vermin and other vectors.
- Keep all storage areas clean and in good repair.
- The manner of storage should maintain the integrity of the containers; prevent the leakage of waste from the container; provide protection from water, rain, and wind; and maintain the waste in a nonputrescent, odorless state.
- Do not use carpets and floor coverings with seams in storage areas. The floor should be impervious to liquids, with a perimeter curve. The floor should be sloped to drains, which are connected to an approved sanitary disposal system.
- Provide the room with exhaust ventilation.
- For security reasons, limit access to the area to those persons specifically designated to manage medical waste.
- Prominently post such areas with the universal biohazard symbol and with warning signs located adjacent to the exterior of entry doors, gates, or walls indicating use of the area for storage of medical waste and denying entry to unauthorized persons.
- Treatment facilities should not store more than seven times the facility's total throughput capacity.
- Maintain storage areas, including vehicles, at a recommended temperature to control odors and to prevent conditions that lead to putrefaction.
- Duration:
  - Minimize storage time.
  - Consider time in transit as time in storage.
  - Storage requirements begin once the container is no longer being filled.
- Collection or transfer stations should submit the following information regarding the medical practice to the state division of solid waste management:
  - Name and address.

- Name of the individual responsible.
- License number.
- Area to be served.

Monitoring and Recordkeeping

Generators of medical waste to be transported to a treatment facility should maintain records of the types of wastes transported, the method of treatment, the designated treatment facility, and the transporter. Such records provide liability protection to the generator as well as documentation of good waste management practice.

Generators who treat waste on-site should keep pertinent records:

- Of spore assay tests.
- Describing the approximate amount of waste treated.
- That demonstrate proper operation of treatment equipment.

Compliance with OSHA Instruction CPL 2-2.44B[6] and other OSHA regulations may require records for each employee, which contain the circumstances of any exposure incident, including date, location, nature of the incident, and the name of the person who is the source of the medical waste. These medical records must be kept confidential, except for disclosure or reporting required by OSHA or by law.

Training

Training of personnel who handle medical waste at the site of generation helps to ensure immediate and accurate identification and segregation of wastes, and safe and effective handling procedures. Training should include the following components:

- Explanation of the waste management plan.
- Assignment of roles and responsibilities for implementation of the plan.
- The epidemiology, modes of transmission, and prevention of HIV and HBV.
- Possible risks to the fetus from HIV, HBV, and other infectious agents.
- The location and proper use of PPE.
- Proper work practices to minimize exposure to infectious agents.
- The meaning of color codes, the biohazard symbol, and precautions that should be followed when handling contaminated articles or medical waste.
- Procedures to follow if a needle stick or other exposure incident occurs.
- Waste minimization procedures.

Implement training when management plans are first developed and instituted, when new employees are hired, and whenever management practices are changed.

Contingency Planning

Medical waste generators should be prepared for unexpected situations, such as accidental spills, loss of containment, exposure, equipment failure, and interruptions or delays in waste collection services that may require the use of alternative facilities. Formulate and disseminate the procedures for handling these incidents to waste handlers.

Keep updated and proper equipment on hand in case of emergency situations involving medical waste. Small quantity generators are likely to produce only certain types of waste and may select only equipment necessary to handle those types of wastes. Larger generators should be prepared to handle larger spills of a wider variety of wastes. The following highlight the steps that should be taken to properly handle unexpected situations:

1. *Management of spills of medical waste.* Keep a spill containment and clean-up kit within the vicinity of any area where medical wastes are managed. The kit must provide information and devices for rapid and efficient clean up of spills anywhere within that area.
2. *Containment and clean-up procedures.* In case of accidental spills, loss of containment, and exposure, the following procedures are recommended:
   - Leave the area until the aerosol settles (no more than a few minutes delay).
   - The clean-up crew dons clean-up outfits and secures the spill area.
   - Spray the broken containers of medical waste with disinfectant.
   - Spray broken containers and spillage inside over-pack bags in the kit, minimizing exposure.
   - Disinfect the area and take other clean-up steps deemed appropriate.
   - Clean and disinfect clean-up outfits before removing.
   - Remove clean-up outfits and place disposable items in clean-up bag.
   - Collect and handle all spill wastes as medical waste.
   - Take necessary steps to replenish containment and clean-up kits with items that were used.
3. *Exceptions for small spills may be allowed.*
4. *Alternative arrangements for waste storage and treatment.* In the event of equipment failure, other forms of storage and treatment should be readily available.
5. *Exposure incidents.*

OSHA defines occupational exposure as reasonably anticipated skin, eye, mucous membrane, or parenteral contact with blood or other potentially infectious materials that may result from the performance of an employee's duties.

OSHA Instruction 2.44B[6] on occupational exposure to bloodborne disease recommends the following procedures to ensure compliance with OSHA inspections:

1. For any persons who are exposed to a medical waste spill, who consents and so desires: collect a blood sample as soon as possible after the exposure incident for the determination of HIV antibody status.
2. Advise the exposed individual to report and seek medical evaluation of any acute illness accompanied by a fever that occurs within 12 weeks of the exposure incident.

3. Offer retesting for HIV antibody to individuals who are seronegative at 6 weeks, 3 months, and 6 months after the exposure incident.
4. Produce a follow-up report on whether any infection due to exposure actually occurred.

Transportation

Medical waste that is not treated at the site of generation must be transported to a treatment or disposal facility. The remainder of waste (eg, treatment residues, treated waste) that has been treated on-site, such as incinerator ash or autoclaved waste, must still be hauled to a permanent disposal facility. Proper management at this state of the treatment process ensures accountability for proper containment and handling of the material until it reaches its final destination.

In the selection of a waste hauler, the generator may want to obtain the following information:

- The size of the hauler's operation.
- State permits or operator's licenses.
- How long the company has been in business.
- Truck type.
- Security measures and worker safety precautions or the details of the waste management plan.
- If the company indemnifies the generator (eg, exempts from loss or damage) for its mishandling of the waste.
- If the hauler provides documentation needed for compliance with federal, state, or local regulations.

Guidelines for transportation are included in The Council of State Governments, *Model Guidelines for State Medical Waste Management*.[7]

Choosing Qualified Disposal Contractors

Many medical practices contract with firms to dispose of their waste. Because the generator retains liability for its waste even after it leaves the facility, the generator should exercise great care in selecting a reputable and experienced disposal contractor. Table 8.4 provides a summary of steps in choosing qualified contractors.

Where should a medical practice begin looking for a contractor? Telephone directories list waste management companies, and state environmental agencies may have lists of firms. Also, check with other facilities and industry groups or associations to determine whom they use and recommend. Insurance companies may also be able to recommend contractors. In addition, waste-handling firms often advertise in trade journals and organizational magazines. Look around and compare prices and services that are offered and screen the companies with a short telephone interview. In a brief conversation, it is often easy to discover much about a company. Look for knowledgeable and articulate answers to questions about regulations, evidence of internal training programs, and the company's programs of compliance.

Screening Potential Disposal Contractors

A prudent first step in screening potential contractors is to check with state hazardous waste management agencies on the firm's background

TABLE 8.4

Summary of Steps in Choosing Qualified Contractors

- Get lists of companies from the telephone directory, environmental agencies, trade journals, and associations to which you belong.
- Conduct a brief telephone interview.
- Judge the company's financial stability.
- Consider the services each firm offers.
- Determine whether the company's insurance coverage is adequate.
- Contact references.
- Obtain descriptions of previous jobs performed.
- Check with environmental officials on the firm's background and compliance history.
- Secure the right documentation.
- Conduct an interview.
- Obtain a proposal, checking for the necessary elements.
- Review the contract for the responsibilities it gives the contractor and any liabilities it may leave for the generator.
- Regularly monitor and stay involved in the contractor's work.

Reproduced with permission from *Managing Health Care Hazards*, Compliance Publication Series. Chaff & Co, 1993.

and history of compliance. They can provide information on how long the company has been operating, whether it has the permits it needs, if any complaints have been filed against the company, or if the company has had any leaks, spills, or environmental accidents. They can also report the results of any inspections they may have made of the firm's facilities and any fines or other means of enforcement used against the company. When consulted, regulatory personnel can act as excellent resources and initiate an ongoing relationship.

A point to consider is the services each firm offers. Some contractors may provide several services, such as waste surveys, technical consulting, and training and reference material. Firms offering this approach can serve as a resource for valuable information and technical assistance.

It is also important to ask about the type of coverage the company has and its limits. Some policies may have exclusions that may leave the generator liable. Look for a company with a solid insurance policy. Along with adequate insurance coverage, the company should also be financially secure. There are a number of resources where this information is available.

The training of vehicle operators is another important consideration. Training should include contingency planning, safety procedures appropriate to the risks they may encounter, how to recognize unacceptable waste, how to recognize a complete and correct bill of lading, and how to use the recordkeeping system used by their company.

Furthermore, a qualified contractor does not hesitate to offer references. Ask for names of organizations for which the company has done work. Be suspicious if the firm refuses to cite a previous job as a reference. Look into it and find out why. It is a good idea to follow up on at least two references, asking frank questions about their satisfaction with the contractor's work and what kinds of problems, if any, have arisen.

Although not mandatory, involvement with a waste management association shows concern for compliance and quality service.

Conducting an Interview

It is important to discuss questions that may arise after reviewing the documentation and contacting references. The following are examples of questions to ask:

- What is the cost of the services?
- What is the cost per month based on a specified average quantity generated?
- If an in-house incinerator breaks down, can the company temporarily handle those wastes?
- Can the company dispose of incinerator ashes?

Obtaining a Proposal

The next step in choosing a qualified disposal contractor is to obtain a proposal from bidders. The proposal should thoroughly explain the type of work to be done, including a regular schedule of services. The proposal should also include the following:

- Describe procedures for transportation, storage, treatment, and disposal.
- Propose safety measures that protect the staff of the facility, the contractor, and the surrounding community.
- Contain an emergency plan that outlines a formalized response to any accidents, spills, or leaks that occur while the contractor is performing work for the facility.
- Provide guidelines for preparing the waste to be transported and documentation of the specifications for disposal containers or bags if they are to be provided.
- Include a cost breakdown, with a maximum cost. (This step prevents hidden costs.).

Drawing Up a Contract

A legal contract must be drawn up that defines the services that the contractor provides and ensures that regulatory, safety, and other requirements are met. The waste firm probably has a standard contract. Review the contract to be sure that it includes the following:

- Assurance of appropriate insurance coverage.
- Schedule of services.
- Cost guarantees.
- Requirements that the contractor complies with all applicable local, state, and federal regulations.
- A section outlining warranties.
- Identification of personnel who will actually perform the work.
- Limitations on changes in the type of waste accepted and services provided.
- Clauses allowing the medical practice to terminate the contract under specific conditions (eg, if the work is not properly performed).

Even with close scrutiny, liabilities hidden in legal jargon can be overlooked. It may be prudent to seek the advice of legal counsel before signing the contract.

After Signing the Contract
The generator's responsibilities do not end with the signing of the contract. The facility should regularly monitor the contractor's work and discuss any problems. Medical practice personnel should stay abreast of new regulations that may affect the waste company's responsibilities. If the facility demonstrates to the contractor that it is informed and interested, regulations and contract requirements are less likely to be violated. The concern and involvement of the generator can mean the difference between increased liability and a job well done.

ENDNOTES

1. Nuclear Regulatory Commission (NRC); Rockville, MD.
2. US Environmental Protection Agency (EPA). *Understanding the Hazardous Waste Rules: A Handbook for Small Businesses—1996 Update*. Washington, DC: EPA headquarters library maintains reference materials and information. Regulatory requirements for small quantity generators are contained in the *Code of Federal Regulations* (CFR), Parts 261 and 262. The EPA Library can refer to state and regional EPA offices.
3. Occupational Safety and Health Administration. OSHA Technical Manual, Section VI, Chapter 2: *Controlling Occupational Exposure to Hazardous Drugs*. Available at: www.osha.gov.
4. US Environmental Protection Agency. *RCRA Subtitle C—Managing Hazardous Waste*. Washington, DC: EPA headquarters library, telephone 202 260-5921 or 5922, maintains reference materials and information. Regulatory requirements for small quantity generators are contained in the *Code of Federal Regulations* (CFR), Parts 261 and 262. The EPA library can refer to state and regional EPA offices.
5. The Council of State Governments. *Model Guidelines for State Medical Waste Management*. Lexington, KY. No. 34-01-033. CAT ACCN 30263. 1992.
6. US Department of Labor, Assistant Secretary for Occupational Safety and Health. *OSHA Instruction CPL 2-2.44B. Enforcement Procedures for Occupational Exposure to Hepatitis B Virus (HBV) and Human Immunodeficiency Virus (HIV)*, Washington, DC: Office of Health Compliance Assistance, February 27, 1990.
7. The Council of State Governments. *Model Guidelines for State Medical Waste Management*. Lexington, KY. No. 34-01-033. CAT ACCN 30263. 1992.

chapter 9

Promoting and Securing Employee Involvement

Go to the people
Learn from them
Love them
Start with what they know
Build on what they have
But the best of leaders
When their task is accomplished
Their work is done
The people will remark
"We have done it ourselves."

Two-Thousand-Year-Old Chinese Poem[1]

Management has a duty and responsibility for "providing a safe and healthful workplace," and each employee has a duty and responsibility to perform in a "safe and healthful" way. Employers and employees are, in fact, partners in this effort. Both reap the benefits of that partnership.

It is not enough for an employer to say:

> I have the equipment and have provided an opportunity for training but no one has to learn to properly use the equipment or to take time for the training. By supplying the equipment, I've done all I can do. Besides, I don't have the time to nag about safe work practices.

Nor is it reasonable for an employee to say:

> Safety glasses make me look silly and are uncomfortable. Those gloves are hot and awkward to work with. Besides, I am too busy taking care of patients, and my boss doesn't care anyway.

There may be a general tendency for people who work in an office environment to disregard safe work practices simply because they think that occupational injuries and illnesses occur only in factories. Most people now work in office environments and a significant number of those are health care facilities. Back injuries among nurses are common, as is sensitization to many of the chemicals that are routinely found in virtually every medical practice.

A third component that is generally not discussed in the same context as occupational safety and health is personal safety and health. People who do not have good personal health and safety habits off the job bring that same attitude to work. Further, illnesses due to poor personal health habits can affect the entire office. Less obvious is the effect of an employee who suffers an injury as a consequence of an unsafe work practice at home. For example, while a person may reluctantly wear respiratory protection at work they may choose not to do so when, for example, they are working with solvents in an unventilated space at home or fail to use proper lifting techniques doing household tasks.

If employers and employees begin to understand the purpose and benefits of safe and healthful work practices, they will extend those practices into their personal life to the benefit of themselves and their families.

What can be done to develop this partnership? There are several steps that can be taken:

- Establish safety and health committees and actively seek the input of all employees.
- Encourage, support, and provide training. An employee's time learning may prevent an injury to someone else and creates an expert.
- Demonstrate management support.
- Design and implement meaningful safety and health promotion campaigns.
- Work with other medical practices to discuss and solve common problems. Involve all practices in area-wide health and safety promotional activities.
- Establish meaningful recognition programs within the medical practice and on a city- or area-wide basis.
- Take every opportunity to relate safe work practices to personal health and safety by promoting the concept among employees and their families.

ESTABLISH MEANINGFUL SAFETY AND HEALTH COMMITTEES

Developing an appropriate and useful safety and health program for a medical practice requires the involvement of a committee that represents employees and supervisors from all departments. Such involvement is essential because employees frequently observe real and potential hazards that supervisory staff or other personnel do not recognize. To be effective, committee members should be knowledgeable in occupational safety and health and have specific responsibilities and appropriate authorities.

During safety committee meetings, it is vital that attendees are:

- Encouraged to participate candidly, with no judgment or reprisal.
- Listened to for their opinion.
- Given every possible opportunity to speak.
- Complimented for good ideas.

Invite suggestions for improved safety practices. Are there any tasks with which employees feel uncomfortable?

The person conducting the meeting should serve as a facilitator rather than the leader. A facilitator makes certain that everyone who wishes to speak has the opportunity and encourages open discussion among the attendees for how best to proceed. Facilitators make sure discussions stay on track and that appropriate amounts of time are spent on issues.

The facilitator also ensures that information is reported through safety committee minutes and oversees the process for acting on reported hazards. If considered an immediate hazard, prompt action should be taken. If considered a long-term concern, document and track the situation.

Action taken through the safety committee should be communicated to employees so that they know their input is welcome and acted upon. This process demonstrates that communication is a two-way street.

Major functions of safety and health committees include:

- Assessing the workplace to identify safety and health hazards.
- Reviewing accident rates, results from prevention activities, and other relevant workplace data.
- Preparing information for employees on identified hazards.
- Organizing educational classes.
- Reviewing safety and health aspects when planning new construction or renovating.
- Investigating incidents.
- Establishing promotional and motivational programs to stimulate employee participation in safety and health activities.
- Evaluating the safety and health program.

Strong and effective safety and health committees require the full support and commitment of management. Committee functions should be a regular part of their job responsibilities.

Joint Labor–Management Committees

Medical practices that are represented by unions should establish joint labor–management safety and health committees. Important contributions can be gleaned by working together to identify problems and by educating the workforce about safety and health issues. Labor unions have played important roles in articulating employee concerns by identifying potential hazards, educating their members, and improving work practices.

INTEGRATE PERSONAL AND FAMILY HEALTH

There are many ways to ensure a safe and healthful workplace, and many of those ways include incorporating personal and family health. The book, *In the Zone*, offers the following methodology for making a difference in everything relating to the body and mind.[2]

Holistic Approach to Injury Prevention

Strains and sprains are the albatross of the American workforce and the national Workers' Compensation System. Back pain is the most prevalent medical disorder in industrialized societies. It disables more than 5 million Americans annually at a cost of $60 billion. The number of health care workers who suffer back injuries constitutes a significant proportion of all work-related back injuries. To begin to correct the problem of work-related back injuries, an overview of how humans interact with all aspects of their environment is required.

Traditional injury prevention programs have been limited in scope, principally focusing on sprains and strains. For example, centering on injury prevention during patient transfer fails to address many other significant factors. Stress, tension, physical strength and flexibility, correct body movement, proper breathing, and mental focus during any form of physical exertion are critical to all injury prevention.

Basic Principles of Body Movement
One of the very first things that every aspiring athlete learns is how to breathe. Physical activity requires exertion. Well-trained athletes understand that to achieve maximum performance they must inhale deeply before an exertion and exhale during the exertion. This simple breathing technique is as relevant to a nurse moving a patient as it is to a weight lifter.

Exhaling during exertion activates the abdominal muscles, thereby protecting the lower back. This basic principle of correct, full-body movement (Principle 1: The human body moves best as a coordinated, integrated unit.) can be easily applied at work, home, and during recreational activities. The second basic principle (Principle 2: Peak efficiency is achieved when the body is relaxed.) of correct body movement can easily be included in a comprehensive safety and health training program. These basic principles of proper body mechanics are highlighted in Table 9.1. Learning how to relax and exert at the same time is key to both preventing injuries and stress management. Understanding how stress can affect the human body is essential to a successful safety and health training program.

Integration of Health Promotion into Occupational Health and Safety Training
Health promotion and occupational health and safety training overlap. If basic occupational safety and health principles are properly taught, the course of instruction shows how those principles apply to an employee's

TABLE 9.1

Two Basic Principles of Proper Body Mechanics

---

1. The human body moves best as a coordinated, integrated unit.
   - Full-body movements dominate in sports, dance, and numerous other forms of human movement.
   - Physical exercise results from "correct" body movements.
   - Repetitive motion injuries result from "incorrect" body movements.
   - Learning how to correctly perform basic human activities can help eliminate injuries caused by strains, sprains, and overexertion.
2. Peak efficiency is achieved when the body is relaxed.
   - Relaxed-body movements are more energy efficient, powerful, and effective than tense-body movements.
   - When people are relaxed and balanced, they notice incremental increases in physical strength and performance effectiveness.
   - When people are tense, out of balance, and distracted, they tend to hold their breath during exertion, which reduces physical strength and increases the risk of injury.

Reproduced with permission from *In the Zone*, Great Ocean Publishers. Copyright 1995.

professional environment as well as personal life. Certainly an employee who is relaxed at work has a far greater tendency to be relaxed at home, thus increasing the possibility of applying the health and safety techniques from work to similar situations at home.

The Total Safety and Health Program
While some people fear change, others embrace it, particularly when they realize that a positive change in their life results from learning new and better ways of doing old things. As the partnership between employer and employee progresses, the health and well being of all aspects of their lives benefit accordingly. A key step in that progression is the integration of occupational safety and health principles with those of stress management, athletic performance, and the ever-present concepts of basic hygiene and nutrition.

Stress

Stress is the body's natural response to external events. Something as simple as a change in the weather or an event far more complex within a person's life can cause stress. Situations that are unfamiliar or challenging increase an individual's level of stress. The level of stress that is experienced depends on personality, state of health, and many other factors. Stress is not always a bad thing. It is a consequence of life. Work, family responsibilities, births, deaths, weddings, and holidays are all reasons that people experience stress. Understanding how an individual responds to stress is key to understanding how they learn and react to change. Some people are motivated by stress and can get great deeds accomplished in that state, which gives them a feeling of achievement. But in others, stress can be debilitating. It increases muscle tension, causes them to be distracted, and makes them almost incapable of functioning. Following are causes and types of stress in personal lives[3]:

- Emotional stress due to arguments, disagreements, and changes in personal life
- Illness or injury
- Overexertion
- Environmental factors
- Tobacco
- Hormonal changes and fluctuations
- Inappropriately assuming responsibility for another's actions
- Allergic responses

The Effects of Stress
People respond to stress in different ways and stressors can trigger different types of responses—some may be physical, some psychological, and some behavioral. If an individual tends to have an anxious personality and low self-esteem then that person is less likely to be able to control the response to stress than someone who has a positive, higher level of self-esteem.

When faced with a stressful situation, a person's initial response depends upon that individual's belief that the skills, abilities, and resources are available to solve the problem. People are also more likely to respond well to these situations if they have good social support from colleagues, family, and friends.

Physiologically, the body responds to stress by increasing its state of arousal in preparation for action. Increased heart rate, blood pressure, muscle tension, and general physical and mental alertness mark this increased state of readiness.

If the stress continues, the body tries to adjust. During this physiological adjustment, hormone levels and heart rates return to normal. If the body's attempt to adjust to the stress is unsuccessful, the individual may experience exhaustion. These rapid energy expenditures tend to reduce the body's ability to successfully respond to attacks by disease-causing organisms. Thus, people who are experiencing a period of significant stress are more prone to illness and disease.

Stress can also manifest itself in behavioral changes, such as increased smoking, alcohol and drug use, and poor work performance. Psychological changes include increased anxiety, depression, irritability, aggression, and sleep disturbances.

Stress in the Workplace

Stress occurs in all workplaces and, as noted above, can have both good and bad effects on individuals, their work performance, and their health and well-being. Efforts to control or manage stress in the workplace should focus on changing the work environment or providing affected employees with help to reduce high levels of stress.

The range of physical, psychological, and behavioral symptoms employees can experience if faced with high levels of stress in the workplace are the same as those that result from stress in their personal life. Some early warning signs are listed below[4]:

- The most common physical effects of stress in the workplace include increased blood pressure, increased heart rate, increased muscle tension, and headaches.
- Psychological effects include increased anxiety, depression, aggression, and confusion.
- The more prevalent behavioral effects include increased smoking, increased drinking, irritability, obsessive concern with trivial issues, and poor work performance. In the workplace, these problems can lead to absenteeism, increased accident rates, poor or reduced work output, and poor interpersonal relations.

There can be a number of causes of stress in the workplace,[4] including:

- Lack of control regarding workloads, overdemanding workloads or tight schedules.
- Lack of clear direction from management.
- Lack of information on work role and objectives, career opportunities, or job security.
- Conflict between individuals or areas, either section rivalry or personal discrimination or harassment.
- Poor physical working conditions (eg, extreme changes in temperature, working conditions that are too cold or too hot, excessive noise or vibration levels).
- Concerns about exposure to hazardous chemicals or situations.

Solutions to Stress

Employers and employees can work together to reduce the level of stress in the workplace.

The following are some simple ways to reduce stress levels at work or at home[5]:

- Determine the nature of the stress.
- Make life regular. People who suffer from stress have often disrupted their biological clocks.
- Give the body a chance to heal following exertion or illness.
- Do not overcommit.
- Explore positive changes in nutrition.
- Avoid allergens.
- Exercise.
- Rest and entertain the mind. Dance, listen to music, read, play an instrument, or meditate.

DEVELOP PROMOTION AND RECOGNITION PROGRAMS

Once a total safety and health program is in place, implement promotion and recognition programs to help reach safety objectives. These programs should serve as the cornerstone of the medical practice's safety philosophy. The concept provided below builds a broad-based program using the best elements of motivational and communication programs from around the world. The program can be anchored in a symbol, logo, or umbrella theme. Suggestions for programs include promotional programs, recognition programs, reward programs, and family safety and health programs. Establish budgets for the programs, which can be implemented in phases with items to fit the budget.

Promotional Programs

Promotional programs increase awareness and expand interest and involvement in safety. They are the commercials of safety and health. Structure promotional programs to effect changes in safety and health practices. Aim the promotions at inducing employees to get involved in educational sessions that highlight safety and health in the workplace. Examples of promotional programs can include:

- In-house TV monitors with safety messages or video interviews with employees.
- Safety hot line.
- Promotional products, such as golf shirts, caps, and T-shirts.
- Annual safety planning meetings with kickoffs.
- Safety suggestion programs.
- Community-wide programs to gain publicity and to generate good public relations.
- National publicity programs through articles in association magazines and training institutes.
- Campaigns for families to vacation safely. Winners could be announced on local radio stations and in the medical practice newsletter.

Recognition Programs

Recognition programs are individualized safety programs that immediately recognize employees for specific behaviors. Examples of recognition programs are as follows:

- Safety pins
- Membership in national societies such as the Wise Owl Club (sponsored by Prevent Blindness America)
- Certificates
- Team trophies
- Newsletter publicity

Reward Programs

Reward programs promote a healthy competition among employees. These programs recognize the efforts of all employees who have met the safety goals of the medical practice. One incentive built into the programs requires that a group, rather than an individual, have a quality performance. Examples of reward programs are:

- Competing toward a department goal.
- Competing toward a company-wide goal.
- Winners of individual medical practices compete with other medical practices for an annual award, such as an invitational golf tournament or a sporting event.

Family Safety and Health Programs

Family safety involvement can be key in reducing employee off-the-job injuries. Family safety and health programs encourage employees and their families to follow the same good safety practices in their outside activities as they use on the job. Figure 9.1 is an example of a family safety activity program depicting a scene from the "Rivera the Cheetah" game on the snappy safari website, available at www.mnsc.org/snappy. Examples of other family safety programs include:

- Vacation safety
- Home fire safety
- Educating children who stay home alone
- Promoting seatbelts
- Electrical safety
- Stories and pictures of family safety news in newsletters
- Utilizing local recreation departments to promote swimming classes or courses in boating safety
- Sending family safety and health materials to employees and their families

SET UP CONTROL METHODS

As people benefit from working safely, they are far more likely to bring their new skills into their personal life. For example, if a person is required to wash one's hands before seeing patients, that routine will become more prevalent at home with family members as they wash their

FIGURE 9.1

An Example of a Family Safety Program Depicting a Scene from the "Rivera the Cheetah" Game on the Snappy Safari Website
Reproduced with permission from Snappy Safari Website, Minnesota Safety Council. Copyright 2000. For more information visit www.msnc.org/snappy.

hands to prevent the spread of disease. Likewise, if a person is trained in proper lifting techniques at work, it is far more likely that those techniques will also be used at home. In other words, once the basic elements of occupational safety and health are learned, those elements are easily adapted to personal safety and health.

Control methods are the primary means of reducing exposure to occupational hazards and include substitution, engineering controls, work practices, PPE, and administrative controls. Employees can also apply these techniques at home while doing routine household tasks. The following sections include fundamental and easily implemented controls for work and home.

Substitution

The best way to prevent occupational safety and health problems is to replace the offending agent or hazard with something less hazardous. For example, in the medical practice, substitute safe needles and syringes to protect employees from potentially lethal needlesticks.

This same concept has value at home. For example, if a certain cleaning agent adversely affects a family member, substitute one that presents a lower lever of toxicity. Vinegar and water, ammonia and water, or rubbing alcohol are effective and inexpensive cleaning agents. Substituting those materials for more exotic cleaning agents presents a much lower level of toxicity and saves a significant amount of money.

Engineering Controls

Engineering controls may involve modifying the workplace or equipment to reduce or eliminate employee exposures. Such modifications include general and local exhaust ventilation, isolating patients or work processes from the hazard, enclosing equipment or work processes, and altering equipment.

Engineering controls used at home include the use of a fan and an open window to remove fumes. For example, if using some type of solvent or irritating chemical, place a small fan behind the person doing the work so that it blows fumes away from the individual and toward an open window or consider doing the work outdoors. In most cases, typical household ventilation is not designed nor is it adequate to efficiently remove chemical fumes.

Work Practices

How employees carry out their tasks may create hazards for themselves and others. For example, staff who do not dispose of used needles safely create a severe hazard for other employees, including housekeepers, laundry workers, and themselves. Employees sometimes perform tasks in ways that create unnecessary exposures. This includes staff members who try to lift patients without assistance or laboratory workers who pipette by mouth rather than by rubber bulb, thereby increasing their risk of injury or contamination.

The same consideration should be given to tasks that are performed at home. Always dispose of broken glass, nails, and lids from cans properly and use correct lifting techniques for moving furniture.

PPE

PPE includes gloves, goggles, aprons, respirators (not surgical masks), and shoe covers. Although the use of such equipment is generally the least desirable way to control workplace hazards because it places the burden of protection on the worker, the equipment should be available for situations when an unexpected exposure to chemical substances, physical agents, or biologic materials could have serious consequences.

PPE is also available for home use. Hardware stores generally sell NIOSH-certified respirators that are effective for exposure to organic vapors, such as those found in paint, and a wide variety of other safety equipment, including dust masks, butyl rubber gloves, and ear plugs. These personal protective devices have a significant role in personal safety and health.

Administrative Controls

Administrative controls involve reducing total daily exposure by removing the worker from the hazardous area for a period of time. Use these controls when it is impractical to reduce exposure levels in the workplace through engineering controls. Administrative controls include rescheduling work to reduce the necessity of rotating shifts and increasing the frequency of eye rest periods for persons who work on computers.

At home, the primary administrative control is ensuring that only those wearing appropriate PPE are in an area where hazardous work is being performed and that those performing the work understand how to do so safely.

ENDNOTES

1. SantoPietro N. *Feng Shui: Harmony by Design*. New York: The Berkeley Publishing Group; 1996.
2. Mulry, Ray. *In the Zone*. Arlington, VA: Great Ocean Publishers, Inc; 1995.
3. *Causes and Types of Stress*. September 27, 1998. Available at: www.muhs.acsu.kiz.vt.us.
4. National Occupational Health and Safety Commission. *NOHSC: Stress and Burnout at Work*. Developed in conjunction with the Essentials, a commercial publishing company. Commonwealth of Australia, www.worksafe.gov.au.
5. *Here are Six Simple Ways to Reduce your Stress Level*. September 27, 1998. Available at: www.muhs.acsu.kiz.vt.us.

chapter 10

Ergonomics

Ergonomics began in the year 1717 when Ramazzini, the "father" of occupational medicine, observed cobblers and craftsmen at work. He determined that many of their injuries resulted from working in awkward positions with irregular motions.[1]

Work-related musculoskeletal disorders impose an enormous burden on workers, employers, and society at large. OSHA reports that musculoskeletal disorders rank second only to cardiovascular disease as a cause of disability in the United States in total economic cost, including direct medical costs and indirect costs due to lost earnings and productivity.

Risk factors can be minimized by evaluating and addressing musculoskeletal problems in the workplace. The medical practice will benefit by reducing employee pain and suffering and decreasing medical, compensation, and other business costs. Even small changes can make big differences in the morale of employees, which contributes to the overall success of the medical practice.

WHAT IS ERGONOMICS?

How employees perform their tasks at work so that they can work in the safest and most efficient manner is known as ergonomics. It also is the study of the relationship between people and their environment, especially in the workplace. Ergonomics covers all aspects of a job, from the physical stresses it places on joints, muscles, nerves, tendons, bones, and the like, to environmental factors that can affect hearing, vision, and general comfort and health. Various types of physical and environmental stress can include some of the following:

- Repetitive twisting movements, usually in combination with poor body position.
- Excessive standing, with no chance to lean, sit, or comfortably reposition the body.
- Repetitive motions using a bent wrist.
- Working in an awkward position, such as holding a telephone to an ear with a shoulder.
- Poor air quality, which may cause headaches, congestion, or fatigue.
- Improper lighting, which can cause eyestrain and headaches, especially in conjunction with a computer monitor.

OSHA provides the following explanation of ergonomics[2]:

> Ergonomics is the science of fitting the job to the worker. When there is a mismatch between the physical requirements of the job and the physical capacity of the worker, work-related musculoskeletal disorders (WMSDs) can result. Workers who must repeat the same motion throughout their workday, who must do their work in an awkward position, who must use a great deal of force to perform their jobs, who must repeatedly lift heavy objects, or who face a combination of these risk factors, are the most likely to develop WMSDs.

People often think of work ergonomics in terms of construction workers, people who operate jackhammers, or other occupations that link tools and equipment to the physical labor of the employee. However, the scope of application is broad. Musicians, typists, artists, athletes, nurses, housepainters, repair workers, dancers, or just about any career involves repeated motions in its performance. Some careers involve greater physical strain and repetitive motion than others.

Ergonomics has certain commonplace terms. Sources often refer to ergonomic injuries as cumulative trauma disorders (CTDs), repetitive stress injuries (RSIs), or musculoskeletal disorders (MSDs). Table 10.1 provides a number of other glossary terms in the field of ergonomics. Ergonomics is an evolving field of study and many terms are not yet standardized. NIOSH and OSHA do not differentiate between RSIs or MSDs. In OSHA's and NIOSH's glossary, MSDs include injuries and disorders of the muscles, nerves, tendons, ligaments, joints, cartilage, and spinal discs. For ease of reference, and because both OSHA and NIOSH use MSD or WMSD as the umbrella term, this chapter uses MSD or WMSD.

The Bureau of Labor Statistics (BLS) records data describing MSDs as "injuries and illnesses" because it defines certain ergonomic problems as an illness (eg, carpel tunnel syndrome). This chapter references the BLS terminology, as appropriate.

TABLE 10.1

Ergonomic Injuries Terminology Glossary

Term	Definition
CTD	Cumulative trauma disorder—synonym for RSI
CTS	Carpal tunnel syndrome
MSD	Musculoskeletal disorders—a less area-specific term for WRULD
OOS	Occupational overuse syndrome—synonym for RSI
RSI	Repetitive strain (or stress) injury—a generic phrase for many types of injuries
WMSD	Work-related musculoskeletal disorders—similar to RSI but restricting the disorder to job related (as opposed to a hobby)
WRULD	Work-related upper limb disorders—again, similar to an RSI but restricting the injuries to the arms

OVERVIEW OF THE PROBLEM

Exposure to ergonomic risk factors on the job leads to MSDs of the upper extremities, back, and lower extremities. WMSDs currently account for one third of all occupational injuries and illnesses reported to the BLS by employers every year.[3] In 1999, this statistic represented nearly 600,000 MSDs that are serious enough to cause an employee to take time off from work. Evidence suggests that an even larger number of nonlost worktime MSDs occur in the workplaces every year. In 1997, US workers reported a total of 626,000 lost worktime MSDs to the BLS, and these disorders accounted for $1 of every $3 spent for workers' compensation in that year. This means that employers are annually paying more than $15 billion in workers' compensation costs for these disorders. Other expenses associated with WMSDs, such as the costs of training new workers, may increase this total to $45 billion a year.

Workers with severe MSDs often face permanent disability that prevents them from returning to their jobs or handling simple, everyday tasks, like combing their hair, picking up a baby, or pushing a shopping cart.

Workers who must undergo surgery for work-related carpal tunnel syndrome (CTS) often lose 6 months or more of work. By the end of 1999, about 1.8 million workers suffered musculoskeletal disorders, and more than 620,000 missed some work because of MSDs.

Musculoskeletal disorders are prevalent in traditional womens' jobs around the world. In a medical office, it is just as likely for front office staff to suffer from an ergonomic injury as office nurses. Office staff may stand for long periods of time, look from a desk to a computer and back numerous times an hour, and put strain on their backs from lifting heavy items in storerooms. Nurses, who help patients onto and off the examination table or help them in and out of a wheelchair, may put repeated stress on their backs and muscles. Nursing jobs also usually entail considerable walking, which may also cause back and muscle stress.

OVERVIEW OF ERGONOMICS IN THE WORKPLACE

When employees are overtly injured on the job, the cause is usually known. The nature of the trauma (eg, broken arm, twisted ankle) may require an X ray to determine the severity of the injury, but the part of the body that is injured is known. The difference between on-site, overt injuries and ergonomic-related injuries or illnesses lies in immediacy. Overt workplace injuries are generally not recognized as "ergonomic-type" injuries, injuries that deal with the long-term *cumulative* effects of performing a daily task. The injurious results may not surface for years, possibly even decades. The purpose of analyzing tasks is to identify ways to prevent cumulative physical trauma.

When there is a mismatch between the physical *requirements* of the job and the physical *capability* of the worker, MSDs may occur. Sometimes MSDs can be reduced by such simple measures as varying the tasks among workers. If two people are required to be seated all day, but two other employees work on their feet (and are qualified and trained in performing the other's duties), having them trade off during the day can greatly reduce the possibility of MSDs.

Verifying an employee's ability to lift items beyond a certain weight can also reduce cumulative stress factors involving the muscles in the back, arms, and legs. The ability to perform a task without risking an MSD is often taken for granted. Assuming that all tasks are being done safely from an ergonomics perspective is perhaps the biggest risk that is taken for granted. A task may be considered safe from a general safety point of view. However, if the worker does the task repeatedly throughout the day, using the same movements, then it is not necessarily safe from an ergonomics viewpoint because the ill effects may not surface for years or decades.

In a medical practice, there are several tasks that can be considered a challenge to proper workplace ergonomics. It is easy to understand that nurses are affected. However, what may not have been considered are the tasks of nonmedical personnel (eg, bookkeepers, insurance claims processors) who remain seated in front of a display terminal for protracted periods.

APPLYING ERGONOMICS IN A MEDICAL OFFICE

Many employers have developed effective ergonomics programs and commonsense solutions to address MSDs in their workplaces. Often MSDs can be prevented by simple and inexpensive changes. For example, adjusting the height of working surfaces, varying tasks for workers, reducing the size of items workers must lift, providing lifting equipment, and encouraging multiple short rest breaks are some simple and inexpensive ways to reduce MSD risks.

For medical office workers who spend a great deal of time on a computer, NIOSH recommends mini-breaks to rest the eyes. The purpose of frequent eye breaks (ie, leaving the work area, performing some other type of work that does not require using a computer, alternating tasks) is to reduce eyestrain and to limit exposure to the computer monitor.

Another benefit of providing a variety of tasks to employees who work all day at a computer is that they have a chance to change the movement and positioning of their hands and arms. Changing the positioning of the hands and arms, providing wrist supports for employees who type all day, and bringing the keyboard to a level that keeps the forearms parallel with the floor are work practices that can help avert carpal tunnel syndrome.

Filing medical records all day, every day, may eventually lead to ergonomics-type injuries. However, if the file clerk can trade off with the receptionist for an hour or two per day, each employee can perform diverse tasks that utilize different bodily motions and stress, thus helping to avert ergonomics-type injury.

Shift work is also a consideration. If the medical practice is large enough to run two shifts, the job tasks of both shifts must be evaluated. What is ergonomically sound for the first shift may not apply to the second because different employees are involved. Within the field of ergonomics, "one size fits all" does not exist. Employers have proven that establishing a systematic program to address such issues as repetition, awkward postures, and heavy lifting results in fewer injuries to employees.

Ergonomic Risk Factors and Solutions

Reducing repetitive motions or physical stresses decreases the potential for MSDs. Examples of these types of tasks include frequent long reaches for supplies that are stored too high or too low for comfortable access, sitting for a long period of time in an awkward position, and repeatedly bending or twisting using an awkward position during patient handling. There are often simple solutions. Rearranging supplies on shelves in the store room, taking frequent breaks from stocking shelves, using team lifting practices, and training may help minimize the potential for injury during the performance of repetitive job tasks.

If an employee sits all day, adding a footrest or a stool can greatly relieve the pressure of a desk chair on the thighs, permitting better circulation. For some employees, a footrest also reduces stress on the lower back. For taller employees, adding a lumbar cushion to a chair can provide spinal support to reduce muscle tension or strain. Some employees, especially those with existing back problems, may find that using a *standing seat* (ie, the backless chair where the knees absorb most of the weight) is beneficial. Some surgeons use them during lengthy surgeries.

Once job hazards are identified, corrective action can be taken, often at nominal or no cost. Solutions to repetitive lifting, however, may not be so easy. Employees may frequently perform lifting tasks, and it is imperative that the types of lifting are suitable for the size, weight, and physical leverage an employee brings to the task. Someone who is 5-foot-3-inches tall may be *proportionately* as strong as someone who is 6-foot-2-inches tall, but can't possibly bring the same leverage to a particular lifting task as can the taller person. Identify the type of lifting to be done, how often, and the physical capability of the individual who is doing it (ie, task in relation to employee).

In the medical profession, a common job-related problem is back injury. While almost any physical motion, including bending, twisting, or reaching, when done repeatedly, can lead to MSDs, lifting remains a major concern. Tasks such as pulling, pushing, or stretching can also lead to MSDs if done on a routine basis and if the employee is not physically capable of sustained, repeated motions of this type.

Frequent complaints about physical aches and pains are among the indicators of improper task-to-employee situations, especially if they do not go away after a good night's rest. Ongoing fatigue may be another indication of undue physical stress.

Encourage employees to be vocal and forthright about physical symptoms they notice (ie, both short- or long-term) as soon as they occur. The employee may need training, the task rotated among employees, or the task modified to adjust to the employee's physical capabilities.

Examples of How to Modify Tasks

The best way to obtain information about a task is to ask those who perform it on a routine basis. Asking open-ended questions (ie, those that cannot be answered with a Yes or a No) may lead to the most efficient solutions. For example, "When do you feel muscle strain?" is more productive than "Do you ever feel muscle strain?" Determining when helps focus on the specific task involved and how often the task is done.

The "modification" in this case may be to rotate staff duties for the employee who is not physically suited to the task. For a short person who must frequently reach overhead to lift down a box of supplies, a stepladder might be a simple and inexpensive means of solving the problem if the supplies cannot be moved to a lower shelf.

In the instance of computer monitor glare, the answer may be found directly on the display. What is the source of the glare? It will probably be reflected on the screen. Overhead fluorescent lights are often the culprit or sometimes it is daylight from a nearby window. Seeking reflections on the monitor's display helps identify the source. Repositioning the display or sometimes tilting the screen slightly downward can reduce or eliminate glare.

Training employees on the benefits of preventive action is equally important. An awareness of good ergonomic practices, whether seated, standing, or walking, fosters an atmosphere of interactive participation. If a coworker comments about an ache or other physical problem, a fellow employee can assist in isolating the cause. Sometimes affected employees cannot see the cause as easily in a routine problem as employees with an objective eye. Encouraging an awareness of problems and the candor to discuss them reinforces that management cares about the staff's well-being.

Simple, basic exercises, as shown in Figure 10.1 and Figure 10.2, can be done at a desk or during a brief break. One of the key elements in preventing MSDs is taking short breaks from the task. The break can come in the form of just standing and stretching, taking a short walk through the premises, performing a few exercises, or working on a completely different task altogether. It is the *accumulation* of time spent doing the same task over and over that leads to MSDs.

Anyone who has ever had a foot or a leg go to sleep from being in the same position for too long recognizes the immediate symptoms of physical stress. Ergonomics centers on the possible long-term effects of physical stress. An analogy might be that poor eating habits would not make young people ill right away, but if they continue an inadequate diet, they are likely to develop health problems when older.

Basis of a Program

The same principles that apply to an effective safety and health program apply to a viable ergonomics program, which include recognition, evaluation, and control of signs that may indicate problems. Table 10.2 highlights the elements of a complete ergonomics program. Initially, it may be necessary to enlist the services of a qualified ergonomics consultant before forming an ergonomics team. When the team is formed, include at least one employee and one individual from management. This combination is important, as employees represent the interests of the staff and management's participation shows commitment to a safe and healthy workplace.

The team should concentrate on those tasks that are *likely* to cause CTDs, such as standing for long periods of time, working with wrists bent, or repetitive twisting movements in combination with poor body position. If employees understand that management *wants* to avert MSDs for the good of the staff as well as to make the program a sound manage-

FIGURE 10.1

Basic Exercises That Can Be Done Throughout the Day to Prevent MSDs

Reproduced with permission from American Network Services, Inc, *Watch Your Back! First Aid for Back Pain.* Copyright 1999.

STANDING ROUTINE

1. Deep Breathing in Position of Strength
Stand with your back straight, knees bent, stomach tight, and arms close. Inhale 14 counts and exhale 14 counts.

2. Backward Bend
Slightly bend your knees and place your hands on the small of your back. Slowly bend backward and hold a few seconds. Keep your eyes focused straight ahead to prevent rolling your head back.

3. Upward Reach
Inhale deeply and exhale as you slowly reach toward the ceiling. Hold and bring your hands down slowly to your sides and return to the Position of Strength. Do a backward bend.

4. Standing Hamstring
CAUTION: This exercise is difficult for back patients. Reach only a little bit at a time and stop before you feel pain. Bending your knees as indicated will protect your back. Do this stretch only as directed. Stand in the Position of Strength with your *knees slightly bent*. Slowly bend forward at the waist, and let your head, neck, shoulders, and arms hang freely as you descend into a Position of Weakness. Now *straighten your legs* and hang in *a* relaxed and comfortable manner. Take a deep breath and relax as you slowly exhale, allowing your neck, shoulders, back, and legs to stretch. When you feel your legs begin to tire, *bend your knees, then straighten your back* as you come upright and stand in the Position of Strength. Use your hands on your knees as extra support.

SITTING ROUTINE

1. Upward Reach
Sit in your chair in the Position of Strength with a straight back. Inhale and slowly exhale as you reach to the sky. Bring your hands down slowly to your sides.

2. Shoulder Blade Squeeze
Reach behind your back and clasp your hands. Squeeze your shoulders together as you straighten your arms.

3. Head Roll
In the sitting position, inhale and then exhale as you drop your chin toward your chest. Slowly roll your head so that your ear is leaning toward your shoulder. Relax and hold. Roll center and lift straight up. Inhale and repeat to the other side. DO NOT ROLL YOUR HEAD BACK or you will compress the spinal column.

4. Forward Bend
Sit forward in your chair so that some weight is on your feet and legs. Spread your feet and slowly bend toward the floor. Drop your chin, arms, and shoulders and hang loosely for about 5 seconds. Return to an upright position.

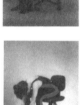

5. Leg Extension
CAUTION: This exercise may be difficult for those with back problems. Stop it if it causes pain. *Inhale* as you extend one leg in front of you and flex your toe toward your head. Exhale as you gently reach forward to stretch the hamstring. Keep your back straight—not rounded. Repeat with the other leg.

ment tool, they are more likely to promptly report symptoms. If the staff also helps to find solutions, opportunities for reducing repetitive stress tasks are increased.

Employees sometimes choose not to promptly report aches and pains, either because they assume the pain will go away or because they do not realize the connection between the discomfort and their jobs. Failing to address the problem and taking steps to correct it may lead to severe

FIGURE 10.2

Stress Prevention Exercises for the Hands

Reproduced with permission from Benchmark Physical Therapy, *Recommended Preventative and Corrective Exercises.* Copyright 2001. For more information visit Website www.benchmarkpt.com. Illustrated by Christopher Chaff, Copyright 2001.

Stress Prevention Exercises for the Hands
Range of Motion Exercises and Carpel Tunnel Syndrome Prevention

Palm Flattening
With hand resting on table palm up, flatten palm by extending fingers and thumb.

 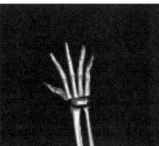

Passive Range of Motion (PROM)
Wrist Flexion/Extension
With elbow bent, hold involved hand and slowly bend wrist forward until a stretch is felt. Relax. Complete by bending in the opposite direction.

 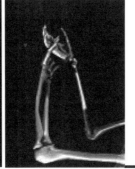

Wrist Flexor Stretch
With elbow straight, slowly bend wrist back with opposite hand until a stretch is felt.

Active Range of Motion (AROM)
Elbow Flexion/Extension (Palm Up)
With palm up, gently bend elbow as far up as possible. Hold. Complete by straightening arm as far as possible. Hold.

TABLE 10.2

Elements of a Complete Ergonomics Program

Program Element	Basic Obligation
Management leadership and employee participation	Demonstrate management leadership of the program. For example, this element would require management to provide time and money to manage the program and implement workplace modifications and training. Employees must have ways to report problems, get responses, and be involved in the program. Do not allow policies or practices that discourage employees from making reports and recommendations or from participating in the program.
Hazard information and reporting	Identify MSDs and MSD hazards in manufacturing and manual handling operations and other jobs that cause MSDs. Provide information about MSDs and their hazards to all employees in jobs. Conduct hazard identification and provide information periodically.
Job hazard analysis and control	Analyze problem jobs. If there are MSD hazards in those jobs, implement measures to eliminate or control the hazards to a feasible extent.
Training	Provide training on the ergonomics program and MSD hazards periodically, but at least every 3 years. The training must be at no cost to employees and must include the following: - Recognition of MSD signs and symptoms and the importance of early reporting. - Reporting of MSD signs, symptoms, and hazards and information on making recommendations. - MSD hazards in employees' jobs and the general measure that must be followed to control those hazards. - Job-specific controls and work practices that have been implemented in employees' jobs. - The ergonomics program and the employees' role in it. - The requirements of the ergonomics standard.
MSD management	Make available prompt and effective medical management whenever an employee has an MSD. This means that when an employee reports signs or symptoms of an MSD, one must check the report to determine whether medical management needs to be provided. Medical management must be provided, including recommended work restrictions, at no cost to the employee. Note: The standard does not contain requirements on specific medical treatments for MSDs. Medical treatment protocols are established by health care professionals.
Program evaluation	Evaluate the ergonomics program and controls periodically, but at least every 3 years, to ensure that they comply with the standard.

Reproduced with permission from *Implementing an Effective Ergonomics Program Flow*, published by *Compliance Magazine*. Copyright 2000. For more information visit www.douglaspublications.com.

disorders later in life. A visual way to explain this is to provide an image of a subcutaneous cyst. It is often assumed the cyst will go away. If it is ignored, it grows and grows until it becomes a tumor. If neglected, the tumor, too, grows. By the time a biopsy is done, should the tumor be malignant, it may be too late. Early intervention applies to numerous situations, including ergonomics.

Safety considerations are different from long-term stress factors. One is not more important than the other, but each must be viewed from a separate perspective. When introducing new equipment, establishing new procedures, or welcoming new employees, examine the ergonomic effects of the changes.

Creating a Receptive Atmosphere

The NIOSH document, *Elements of Ergonomics Programs*,[4] reports that work-related MSDs are often underreported. Therefore, the need to recognize early signs and symptoms of MSDs and to be aware of MSD hazards in the workplace is particularly important. Some reasons for the underreporting of work-related MSDs include:

- The difficulty of linking symptoms, such as pain and tingling, to workplace risk factors.
- The belief that pain is a part of the job and that nothing can be done about it.
- Intentional or unintentional discouragement of reporting by management.
- Employees sometimes fear reprisal for reporting.
- Employees are discouraged from reporting by their supervisors or managers.
- Employees are discouraged from filing a workers' compensation claim.
- Employees want to avoid the hassle of filing a workers' compensation claim.
- Employees prefer (or are encouraged by their employers) to use the employer's or their own health insurance rather than the workers' compensation insurance system.

A receptive and committed atmosphere that promotes employee participation in the program and encourages them to report MSDs is important to the success of the program. Involving the entire staff in finding job hazard solutions strengthens the program and helps employees understand that the effort is critical to their well being. Even if an employee changes career direction several years' later, cumulative effects of a decade of repetitive tasks will not disappear.

When searching out improved ergonomic working conditions, employees must take responsibility for their own future well-being. Employees who continuously slouch, for example, are likely candidates for back problems later in life. Seated employees are ultimately inviting circulatory problems when they wrap their ankles around the base of the chair, cross their legs, or put added pressure on their thighs against the seat of the chair. For employees who stand all day, wearing ill-fitting or inappropriate shoes can also lead to cumulative disorders.

Work habits may sometimes be corrected by briefly chatting with employees. Other times, training on the physiological effect of poor posture or unnatural body positioning while performing certain tasks may be necessary.

As with other safety and health issues, MSD reporting may increase temporarily as the program gains recognition. As employers take action to implement the elements of the program and employees participate in creating a safe work environment, MSDs are prevented and lead to a decrease in reporting and the dwindling of workers' compensation claims.

Identifying Problems

Use a proactive, preventive approach in identifying problems and determining when and what action is needed. Approaches include surveying employees and reviewing reports of MSD symptoms. Medical research indicates that the earlier an MSD is addressed, the more likely it can be reversed.

In a small practice of one or two employees it should not take long to identify specific tasks and review them with the individuals who perform them. A worker may have already commented about needing a stepstool, excessive glare in the workstation, or lack of back support from a desk chair.

Larger practices may require focused time to identify ergonomic needs and solutions. A nurse is not likely to be aware of specific problems that office personnel may have and vice versa.

If the medical practice includes a laboratory, consider a job hazard analysis of the department. If one technician is 5-foot-2-inches tall, but the other is 6-foot-1-inch tall, the height of table surfaces and work stools may pose a problem. Performing tasks for 8 hours a day in an uncomfortable, awkward position can lead to MSDs. However, there are systems available that attach to work tables making them quickly adjustable to suit any employee's height. Modern work stools also come with a means to easily and efficiently adjust the height.

Incident Reports

Incident reports and other data tracking reports are good ways to determine whether or not there are ergonomic problems in the office. Note the frequency of each type of problem and its severity. If most workplace injuries can be traced to a single source, for example, helping patients onto examination tables, conduct a job hazard analysis of that job. The nature of the report, including incident, injury, or workers' compensation claim, is an excellent indicator of what aspects of the practice need an ergonomics evaluation.

Even if there have been no incident reports or workers' compensation claims throughout the past year, check how many workdays have been lost. They may or may not be job related and should be followed up to eliminate the worksite or job factors as the problem. Another possible way to gather information is to observe body language, which often indicates problems with muscles, eyestrain, headaches, and other symptoms that lead to MSD.

Do not make decisions with regard to safe procedures or practices, but confine them to *task*-related-to-*worker*. Table 10.3 illustrates sample job factors to consider when conducting a job hazard analysis. For example, a single employee who is absent several times a year due to a strained back muscle should receive a medical evaluation and perhaps be reassigned. Likewise, if numerous employees complain about shooting pains in one or both wrists, a job hazard analysis should also be performed because repetitive actions may be the cause.

Nurses and doctors are prime candidates for MSDs. They often stretch, lift, or place stress on muscles by working for prolonged periods of time in awkward positions, ultimately leading to a greater cumulative strain.

Management leadership and employee participation in the job hazard analysis results in a better evaluation of where ergonomic improvement is necessary and what priorities need to be set to decrease MSD hazards.

Tasks and Workstation Interrelationships

Many computers or display terminals pose health problems that can be effectively avoided by adjusting the distance between the screen and the user. The position change, however, can have an effect on the employee via eyestrain or headaches.

Certain tasks involving laboratory work could lead to MSDs. If no complaints have been filed, talk with employees and observe both the work area and actual work practices. There are questions to ask employees. For example, do any employees notice back strain? What about sore limbs or muscle aches? Do employees experience a numbness or tingling in their fingers, either on- or off-duty? Do any employees suffer from an unusual number of headaches or frequent eye fatigue? Another means of seeking out potential problems is by paying attention. If employees are squinting or rubbing their temples, it could be a sign of eyestrain. Between observations and asking questions, potential problems can be identified.

X-ray technicians may also develop MSDs. For example, patients sometimes need help to and from their wheelchair. Frequent adjustment of equipment to suit the patient's size or the placement and frequent removal of X-ray plates may also lead to cumulative MSDs.

Employees may complain about nonwork-related aches and pains. Unaccustomed exercise over a weekend, painting the house, or any number of other nonemployment factors may contribute to MSD symptoms. Setting up and managing an effective ergonomics program provides the leadership and reporting mechanisms necessary for accurate problem identification, employee reporting, and MSD prevention.

COST VS. BENEFITS

The consequences of occupational injuries and illnesses of all kinds extend far beyond their immediate, short-term effects. Loss of self-esteem, disruption of family life, and feelings of anger and helplessness are the frequent consequences of any disabling occupational illness or injury. This personal toll cannot truly be captured in monetary terms because the human dimension of occupational injuries and illnesses—

TABLE 10.3

Sample Job Factors to Consider in a Job Hazard Analysis

Physical Work Activities and Conditions	Ergonomic Risk Factors That May Be Present
1. Exerting considerable physical effort to complete a motion	Force Awkward postures Contact stress
2. Doing same motion over and over again	Repetition Force Awkward postures Cold temperatures
3. Performing motions constantly without short pauses or breaks in between	Repetition Force Awkward postures Static postures Contact stress
4. Maintaining same position or posture while performing tasks	Awkward postures Static postures Force Cold temperatures
5. Sitting for a long time	Awkward postures Static postures Contact stress
6. Objects or people moved are heavy	Force Repetition Awkward postures Static postures Contact stress
7. Bending or twisting during manual handling	Force Repetition Awkward postures Static postures

Reproduced from www.osha.gov proposed ergonomics standard.

the pain, suffering, loss of self-esteem, and reduced quality of life—cannot readily be expressed in numerical terms.

Some dimensions of occupational injuries and illnesses can be quantified in monetary terms. Financial aspects of WMSDs can be measured by the lost time experienced by employees and by other costs externalized throughout society. According to OSHA, failure to control jobs that give rise to MSDs accounts for a large part of the costs of expensive income-maintenance programs, such as temporary workers' compensation and permanent disability programs. These costs impose a burden on society that is separate from, and in addition to, the human toll of pain and suffering caused by workplace injuries and illnesses.

With almost any aspect of business, there are direct and indirect costs. Workers' compensation manifests numerous direct and indirect costs. Understanding costs is important when considering how to calculate the cost of implementing an ergonomics program against the benefits of a program.

Direct costs include the lost output of the worker, as measured by the value that the market places on that individual's time. This value is measured as the worker's total wage plus fringe benefits.

Indirect costs describe the costs of work-related injuries borne directly by employers but are not included in workers' compensation claim costs. Examples of indirect costs include payment of sick leave for absences shorter than the workers' compensation waiting period, losses in production associated with the injured worker's return to work, losses in the productivity of other workers, medical expenses, and a wide variety of administrative costs such as MSD management costs for the injured worker. Telephone calls and faxes between the medical office coordinator and the insurance company are also time-consuming and part of indirect costs. If the employee is never able to return to work, time and money are spent interviewing, hiring, and training a new employee.

One example of an ergonomics-related workers' compensation claim is a $29,000 carpal tunnel claim, which resulted from an employee spending long hours typing at a computer. The direct cost in premiums is calculated at $37,400 for 3 years. The total sum adding in the indirect costs is $71,450.

MSDs, such as carpal tunnel claims, can usually be avoided. The most important element of prevention is to reduce tension in the muscles and tendons. This requires learning how to relax, especially when under a lot of stress. Preventive exercises and techniques include job rotation, brief breaks, good posture, and a good night's sleep. Regular exercise and stretching can help strengthen muscles and prevent injury.

Alternative (ergonomic) keyboards are gaining popularity in the workplace. NIOSH cautions that a computer keyboard is only one element in the workplace that influences comfort and health.[5] Examine all features of the computer work environment, not just the keyboard, when evaluating ways to enhance comfort and avoid potential musculoskeletal problems. Table 10.4 provides suggestions from NIOSH for selecting an alternative keyboard.

For some people, wrist pads seem to be beneficial. However, with good posture and the keyboard in the correct spot, the wrists should not be resting on anything. Employees observing each other at work (eg, sitting, typing, and relaxing) may notice more easily if another employee's shoulders are hunched or hands turned.

Another example of an ergonomics-related workers' compensation claim is a lumbar injury with $65,000 in direct costs and another $58,645 in additional premiums over the following 3 years. Once all indirect costs are added, the total cost of the lumbar injury is $131,385. Lumbar injuries can be prevented with training on lifting techniques, placing heavy supplies at the right height, or verifying that an employee is physically fit for the task.

Additional indirect costs often include lowered employee morale. If the staff perceives that the medical practice's management is unconcerned or just too cheap to take preventive action to ensure the staff's well-being, diminished team spirit can lead to poor patient service, increased absenteeism, and possible damage to equipment.

RESOURCES AVAILABLE FOR ONGOING AWARENESS

As the field of ergonomics grows, so do the resources, such as NIOSH documents[6] and mailing lists,[7] that are available to aid employees. Memberships are available in organizations that provide safety and

TABLE 10.4

NIOSH Suggests Considerations for Alternative Keyboard Use

- Determine if the keyboard is compatible with existing hardware and software, and whether it can accommodate other input devices such as mice and trackballs.
- Access how the keyboard will fit with the workstation. Some alternative keyboards, particularly those with a tented design, must be placed on surfaces that are lower than those required for standard keyboards to achieve proper working posture.
- Evaluate whether the keyboard will affect the user's performance. Does the design make it difficult for the user to see the keys? Does the job require a numeric keypad or specialized keys that may not appear on an alternative keyboard?
- Allow users to try a keyboard on a trial basis before buying it.
- Because one type of keyboard will not be appropriate for all users or tasks, allow users to try different kinds before deciding which to buy, and allow them to retain a conventional keyboard if they wish.
- It may take a few days for a user to become accustomed to an alternative keyboard, and frustration may occur if productivity is affected during this learning phase.
- It can be helpful to involve a specialist who knows about and is experienced in office ergonomics, and also to involve a health professional if a computer user has discomfort or musculoskeletal symptoms.
- Integrate a new alternative keyboard carefully into the work process, ensuring that users are trained in correct use.
- Each workplace should have a comprehensive ergonomics program to protect all workers.

Reproduced from NIOSH Publication 97-148.

health information. As more evidence is gathered and newer solutions provided, what might have once been a hopeless problem years ago, may now have a simple and economically feasible solution.

OSHA and a variety of other sources provide much literature and technical expertise on ergonomics to employers. Many publications, informational materials, and training courses are available from OSHA regional and state offices or online at OSHA's ergonomics website (www.osha.gov). OSHA also provides free on-site consultation services to employers who request help in implementing their ergonomics programs.

Some major insurance carriers now offer consultation and seminars, more universities are offering courses on ergonomics, and companies provide consultation. The Internet and safety-related websites are other ways of staying abreast of safety and health breakthroughs. Internet resources help to focus a search on material that applies only to medical offices.

An entire industry has grown out of ergonomics—from air-powered hand tools to height-adjustable chairs, stools, and worktables to practical solutions to preventing MSDs. The study of ergonomics now encompasses manufacturers of office furniture, floor mats, tools and equipment, noise-reducing cubicle enclosures, industrial applications, and a host of other items designed to make the workplace ergonomically sound.

If a practice is new or relocating, consider ergonomically designed office furniture, tools and equipment, and examination room equipment. Purchasing ergonomically correct furniture, fixtures, and equipment can be considerably less expensive than replacing or retrofitting at a later date. Many computer-related problems can also be prevented with adjustable chairs and desk surfaces, as well as accessories, such as nonglare screens for the display and ergonomically redesigned keyboards.

One of the essential tools in preventing MSDs is physical fitness.[8] Ergonomic injuries are less likely to occur if employees maintain good physical health. There are several ways the employer can participate in this effort. One way is to provide information and training on the importance of physical fitness. Other ways include subsidizing a fitness program through a corporate membership or providing an on-site fitness center.

OSHA AND ERGONOMICS

There is considerable debate regarding the need for an OSHA standard concerning ergonomic risk factors in the workplace. Whether or not ergonomics becomes a separate OSHA standard, OSHA still holds workplaces accountable for ergonomics-related injuries and illnesses under the General Duty Clause. If an ergonomics standard is finalized and published, OSHA plans to grandfather in those companies whose programs are in place. Medical practices should consult OSHA regulations for specific compliance requirements.

ENDNOTES

1. Ramazzini, Bernardino. *Diseases of Workers.* Translated from the Latin text DeMorbis Articum (1713) by Wilmer Cave Wright. OH&S Press 1993. NorthWest Training and Development: Thunder Bay, Canada.
2. Occupational Safety and Health Administration (1999). *Ergonomics.* Available at: www.osha.gov.
3. DOL. *News.* US Department of Labor, Bureau of Labor Statistics: Washington, DC. March 28, 2001.
4. National Institute for Occupational Safety and Health. *Elements of Ergonomics Programs.* 1997. US Department of Health and Human Services, Public Health Services, Centers for Disease Control and Prevention, National Institute for Occupational Safety and Health: Cincinnati, OH. DHHS (NIOSH) Publication No. 97-117.
5. National Institute for Occupational Safety and Health. *Alternative Keyboards.* 1997. US Department of Health and Human Services, Public Health Services, Centers for Disease Control and Prevention, National Institute for Occupational Safety and Health: Cincinnati, OH. DHHS (NIOSH) Publication No. 97-148.
6. National Institute for Occupational Safety and Health. *Musculoskeletal Disorders and Workplace Factors: A Critical Review of Epidemiologic Evidence for Work-Related Musculoskeletal Disorders of the Neck, Upper Extremity, and Low Back.* July 1997. Second Printing. US Department of Health and Human Services, Public Health Services, Centers for Disease Control and Prevention, National Institute for Occupational Safety and Health: Cincinnati, OH. DHHS (NIOSH) Publication No. 97-141.
7. National Institute for Occupational Safety and Health, Publications Dissemination; Cincinnati, OH. For more information, visit www.cdc.gov/niosh.
8. Mulry, Ray. *In the Zone.* Arlington, VA: Great Ocean Publishers, Inc; 1995.

chapter 11

Workplace Violence

Administrators or executives who allow violent actions, words, or threats to go unnoticed without discipline are sanctioning violence.

Ann Wolbert Burgess, Examining Violence in the Workplace, **Journal of Psychosocial Nursing, 1994**

While collecting statistics for 1998, NIOSH found that on average 20 employees are murdered every week and that murder was the second leading cause of workplace fatalities. These figures include all types of employment, but the chances of homicide are greatest for those in law enforcement, retail trades, and taxi driving.

RISK FACTORS

Employees in the health and social services, as well as retail positions, face an increased risk of work-related assaults.[1] According to the Justice Statistics' *National Crime Victimization Survey*[2] (a household survey), about two million workplace assaults occur each year.

According to OSHA, heath care and social service workers face an increased risk of work-related assaults stemming from several factors, including:

- The prevalence of handguns and other weapons among patients, their families, or friends.
- The increasing number of acute and chronically mentally ill patients now being released from hospitals without follow-up care who now have the right to refuse medicine.
- The availability of drugs or money at medical practices and other health care facilities, making them likely robbery targets.
- Situational and circumstantial factors, such as unrestricted movement of the public in clinics; the increasing presence of gang members, drug or alcohol abusers, trauma patients, or distraught family members; long waits in emergency or clinic areas, leading to client frustration over an inability to promptly obtain needed services.
- Low staffing levels during times of specific increased activity, such as meal times, visiting times, and when staff are transporting patients.
- Isolated work with clients during examinations or treatment.
- Solo work, often in remote locations, particularly in high-crime settings, with no backup or means of obtaining assistance, such as communication devices or alarm systems.

- Lack of training of staff in recognizing and managing escalating hostile and assaultive behavior.
- Poorly lighted parking areas.

DEFINING WORKPLACE VIOLENCE

OSHA and NIOSH define workplace violence as any act of aggression, physical assault, or threatening or coercive behavior that causes physical or emotional harm and occurs in the workplace. It includes, but is not limited to, the following:

- Beating
- Stabbing
- Suicide
- Shooting
- Rape
- Verbal abuse
- Threats (including bomb scares, obscene phone calls, or even sexual and verbal harassment)

From an anecdotal perspective, it appears sick or injured patients often display one of two attitudes: docile or aggressive. Those who work in the profession must be prepared to deal with the latter.

Suicides occur in the workplace and, while they are not an act of aggression against others, the psychological repercussions can have a profound and lasting effect. Acts of sabotage at the worksite are also a form of violent aggression. If the computer system is disrupted or a bomb is hidden on the premises (even if discovered and defused), the effects upon the staff can be profound.

Types of assault may include pinching, biting, hitting, grabbing, kicking, or striking with a weapon. It may be difficult to imagine what kind of "weapon" could be found in a medical practice, but anything will do. Pencils, ballpoint pens, framed pictures, lamps, even chairs, can all be used as weapons. Someone who is intent on inflicting injury will find a way to hurt others.

The news media often concentrates on sensational instances that result in multiple injuries or death; however, such violent acts represent only a small fraction of the number of workplace violence incidents. Nearly one million people become victims of violent crime while working or on duty. These acts of violence can be committed by visitors, vendors, coworkers, former coworkers, as well as the patients. This results in an annual average cost of 3.5 days of missed work per crime but does not begin to include the emotional and psychological costs to victims, their families, and to the witnesses of the crime. It also does not count the effect on morale, concentration, productivity, or how employees perceive the safety factor within the medical practice.

Some employers fail to report violent crime, believing that the crime is not serious enough or that they can handle it themselves. Others fail to report violent crime because they fear that publicity of such incidents may negatively affect their image in the community. Employers should report threats and violence to authorities and inform their workforce so that they can take appropriate precautions for their own safety in the future. Failure to acknowledge an act of violence—regardless of how minor—can have fatal consequences.

Recordkeeping and program evaluation are essential in determining the overall effectiveness of a workplace violence prevention program. Identification of deficiencies or changes to the prevention program should be noted.

TYPES OF WORKPLACE VIOLENCE

In 1998, 36 homicides of health care industry employees occurred in the workplace, up from 22 such deaths in 1997.[3] According to the Bureau of Justice Statistics' Special Report,[4] and NIOSH's 1996 publication, *Violence in the Workplace*,[5] the statistics for the period between 1992 and 1996 are as follows:

- Health care workers are assaulted in the workplace more frequently than any other US worker group, including police officers.
- More than 160,000 workers in medical fields (including 10,000 physicians and 69,500 nurses) annually reported being the victim of nonfatal assaults in the workplace. Unarmed assailants perpetrated almost 90% of those nonfatal assaults.
- Health care patients were the source of half of all nonfatal assaults against workers that were reported to the US Department of Labor.

Medical practice employees face special considerations in the area of workplace violence. They may work alone with patients in private or isolated areas out of sight or hearing by the rest of the staff. They often deal with patients with illnesses that may impair their judgment and reaction, especially if the procedures are painful or embarrassing.

Patient frustration over billing and insurance issues, such as payment delays or erroneous information, can also contribute to stressful situations that may result in violence. Very often, it is not the office individual the patient perceives as "the enemy," but what that individual represents. For example, if a patient who carries health insurance is being dunned for payment by a medical practice, the practice, not the insurance company, may be blamed.

There are countless reasons that could influence violent behavior. For example, a protracted traffic delay on a summer day means the patient is already out of sorts and perhaps miserably uncomfortable. Similarly, a patient may be having an adverse reaction to prescription drugs or an addict might forcibly enter the premises with intent to steal more drugs. To complicate the matter further, some states now permit carrying concealed weapons as long as the person carrying the weapon obtains a legal permit. Legal or not, the staff in a medical practice is typically busy and might not notice the presence of a weapon.

Geographic and cultural influences can also affect workplace violence issues. Our society has generally become more violent, which increases the risk of being involved in a violent incident. Practices or clinics located in outreach neighborhoods may not have adequate security protection needed to prevent violent incidents. Medical practices that provide family planning and abortion services may also be at a higher risk for violence. For years, the Right to Life Movement has been plagued by the actions of individuals who commit acts of violence. These vigilante actions include bombings, physical harassment, killings, and other acts of violence.

Patient and Staff Interaction

Medical practices face special considerations when developing plans for prevention of workplace violence. The staff is often dealing with patients who, due to illnesses, are more stressed than at other times. Chronic pain, for example, often produces irritability and even hostility. When people are ill or pending a diagnosis or prognosis, they frequently lash out at others due to fear of the unknown.

These outbursts can be verbal or physical. Sometimes outbursts are due to patient dissatisfaction, such as having to wait too long to be seen or taking longer than usual because the practice is short-handed that day. Employees who work in a medical practice are usually exposed to some type of abuse at one time or another.

Suggestions for improving patient and staff interaction include practicing compassion and listening to patients to understand their unique concerns.

Protests

The political issues of health care are often the source for violent or hostile actions that expose health care workers to increased safety risks. Concerns over the use of unborn fetuses to cure certain illnesses, assisted suicide with terminal patients, requiring DNA tests before putting any prisoner to death, or using a baboon's heart for transplant into a human are all hotly debated issues. They will provide medical, theological, legal, and political controversy for years to come. Furthermore, the preliminary mapping of the human genome is already an issue with regard to privacy. Will it give health insurance carriers the ability to accept or deny people for coverage based on their genome? In addition, the Pro-Life and Pro-Choice conflict has led to the death of physicians and their staffs, shootings, bombings of clinics, and many other violent crimes.

If a medical practice is likely to be subject to any group's protest or demonstration, the employees need to know what to do. Some demonstrations are nonviolent, enabling the staff to continue their duties. Violence sometimes erupts from what starts off as a peaceful demonstration. Does the situation require police protection? When arriving or leaving for work, are there any specific safety procedures that employees should follow? By planning for various scenarios, a medical staff will be more prepared to deal with a situation and be able to plan for their own and their patients' safety.

Health clinics—private or government funded—seem to be subject to protests more than private practices. The entire staff should be educated on procedures that are in place for their protection. Equally important, staff members should know their roles in the event of a protest and be aware of any arrangements with the local police department for handling these situations.

It is not uncommon for a news event to trigger a demonstration. Because such events cannot be foreseen, keeping current with local and national news events can aid in determining the risk factor at a clinic or medical practice. The risks at each group practice are different. Each medical practice should undertake an individual risk assessment. This assessment aids the practice in developing their workplace violence prevention program and should include:

- Acceptable conduct by staff in carrying out their duties
- How the staff will be provided with a safe work environment that includes safe passage to and from work
- The extent to which police and media will be involved
- Who the clinic spokesperson will be
- What aftercare counseling is available
- How staff is educated to appropriate procedures

An assessment of the likelihood of protests and demonstrations should be regularly updated, and the procedures should reflect changes in risks.

Bombs and Bomb Threats

Much of the media attention to the problem of workplace violence in health care facilities happens after something dramatic occurs. Abortion and family planning clinics remain high on the list of targets. This might lead the general public to believe that bombings are the most frequent and deadliest problem of workplace violence in medical practices; however, statistics do not support this theory.

Between 1991 and 1998, 7 people were killed and 19 wounded at abortion clinics from bombings or gun shots. In 1998, more than 10% of all abortion clinics nationwide reported that they had received bomb-threat calls.[6] While a bombing is "everyone's worst scenario," in reality people are more likely to lose their lives in a medical practice through an assault by a patient or other person who has legitimate business with the facility than through a bomb. In 1998, there were no reported deaths due to fires or explosions in general group practice health care facilities. While several abortion clinics suffered damage due to bomb explosions, there was only one fatal abortion clinic bombing that left one person dead and one seriously injured.

While many of the abortion clinics nationwide have already increased security measures for staff and patients, continual assessment and upgrading of the workplace violence program are needed to help the staff cope with this risk. Other medical practices could also benefit from developing workplace violence prevention programs to increase staff safety in the event of a violent incident.

Threats to Medical Staff Personnel

Because threats are usually verbal, they are often downgraded to "heat of the moment" comments. Threats are a much more frequent pattern of workplace violence. With the exception of bomb or sexual harassment threats, the majority of incidents are not officially reported.

Some examples of threatening behavior include:

- Intimidation
- Threats of violence
- Stalking

- Verbal harassment, including demeaning or ridiculing
- Violent reactions to disciplinary or other procedures

Threats sometimes occur between a supervisor and an employee, such as during performance reviews. Certain behaviors and language may be perceived as threatening when there is no such intent. Telephone threats from a patient, protester, former employee, or prank caller may occur. Other threats may be the result of a domestic violence situation that spills over into the workplace.

In these instances, take measures to protect both the staff and the person making the threat. In some situations, the threatened person may be face-to-face with the antagonizer; however, if there is a threat to an absent third person, consider contacting that person and alerting law enforcement officials. A situation involving staff threats requires considerations separate from those required in therapist–patient communication (ie, notifying third-party victims or involving potential victims in the assessment and management of the situation). In therapist–patient communication, supreme courts and lower courts often rule differently regarding liability for failing to warn intended victims.[7,8]

Medical practices should develop a checklist or script for specific types of threatening calls, such as bomb threats, suicidal calls, angry calls from patients, and calls to employees involving domestic violence. Checklists can help focus a staff member during the threatening call, get the information that is needed, and provide skills to calm the situation. Figure 11.1 is an example of a checklist that can be used when a medical practice does receive a threatening call. Medical practices should provide ongoing training and education to help staff deal with threats and may find it helpful to incorporate role-playing as part of the training program.

Stalking

In recent years, stalking has emerged as a major social and legal issue. Legal definitions vary widely among states, but most require that the behavior be repeated and that it be harassing or threatening. The definition in a recent survey by the National Institute of Justice (NIJ)[9] defined stalking as "a course of conduct directed at a specific person that involves repeated visual or physical proximity, nonconsensual communication, or verbal, written or implied threats, or a combination thereof, that would cause a reasonable person fear." Stalking is a form of violence sometimes overlooked when developing a workplace violence prevention program because victims may not perceive themselves in harm's way. Patient anger can lead to stalking and often results when patients feel rejected or abandoned. The NIJ survey found that stalking was far more prevalent than anyone had imagined. More than one million Americans are victims of stalkers every year.

The following are forms of stalking behavior that could occur while levels of threat are still low:

- Vandalizing victim's property (eg, car)
- Nuisance telephone calls (eg, hang up, silence)
- Threatening telephone calls
- Hate mail (eg, from newspaper clippings)

FIGURE 11.1

An Example Checklist for Use When a Threatening Call Is Received
Reproduced with permission from Linda Glasson, *Violence Prevention Program*. Copyright 1998.

THREAT CALL CHECKLIST

1. Date _____ Time Call Received _____ Time Call Ended _____

2. Message Received: _____
   _____
   _____
   _____

3. Caller's Voice

   Calm ____ Angry ____ Excited ____ Stutter ____ Lisp ____ Slow ____ Rapid ____
   Soft ____ Raspy ____ Ragged ____ Laughing ____ Crying ____ Normal ____
   Clearing Throat ____ Distant ____ Slurred ____ Nasal ____ Deep Breathing ____
   Cracking Voice ____ Age ____ Disguised ____ Accent ____

   Familiar ____ If so, who did it sound like? _____

4. Background Noises

   Street Noises ____ Animal Noises ____ Office Sounds ____ Clear ____ Home ____
   Static ____ Voices ____ Local Call ____ P.A. System ____ Unknown ____
   Motors ____ Vehicles ____

   Other _____

5. Threatening Language

   Educated ____ Well-spoken ____ Foul ____ Taped ____ Read ____
   Irrational ____ Incoherent ____

   Other _____

6. Number call came in on _____

7. Do you suspect anyone? No ____ Yes ____ If yes, name _____
   Why? _____
   _____

8. Your Name _____ Department _____
   Position _____ Extension _____

Education is a key to learning how to proactively deal with stalking. Staff handbooks and workplace violence prevention programs must cite stalking behavior as unacceptable. Training on conflict resolution is recommended to defuse situations before they escalate and to help identify dangerous situations in the early stages. If stalking is a crime in the community's jurisdiction, list police numbers in the office area to report the problem and document the call. Community resources should also be listed in training documents.

There are many different terms for stalkers. However, stalkers fall into primarily three broad categories: intimate partner stalkers, delusional stalkers, and vengeful stalkers.[10]

Intimate Partner Stalkers

Over half of stalkers fall into this category. Intimate partner stalkers are typically known as the person who "just can't let go." Intimate partner stalking is often an outgrowth or contributing factor of domestic violence that can lead to serious consequences. In an effort to exert control over victims, abusers may even prevent them from going to work. Alternatively, they may show up at work to harass the victim.

From a business perspective, being stalked results in lost time, excessive use of sick leave, and loss of productivity. People are less productive if concerned about what might happen when they leave work. Stalking also negatively affects the workplace because stalkers know they can find their victims at work. At the same time, however, the workplace is an excellent place for an employee to seek help away from the eyes and ears of the abuser. When management is sensitive and responsive to the concerns of employees, victims are much more likely to get the help they need.

Delusional Stalkers

Some studies show that delusional stalkers are the most tenacious of all. What they have in common is some false, idealistic belief that keeps them tied to their victims. Delusional stalkers frequently have had little, if any, contact with their victims. They may have major mental illnesses like schizophrenia, bipolar depression, or erotomania and can be dangerous. For example, the man who stalked actress Rebecca Schaeffer developed an obsession with her and struck her down, and one of singer Madonna's stalkers claimed to be her husband coming home to spend a "frolicsome" weekend.

Vengeful Stalkers

Vengeful stalkers become angry with their victims over some slight, whether real or imagined, and they stalk with the intent to get revenge. Disgruntled ex-employees can stalk, whether targeting their former bosses, coworkers, or the entire medical practice. These people can stalk staff members to and from work without their knowledge. The victims in such cases are often in great danger, for some of these angry stalkers are people without conscience or remorse. They can be delusional (most often paranoid) and believe that they are the victims of society, the system, or simply "them" where "them" is a faceless entity. Bottom line, vengeful stalkers all stalk to get even.

Patients have been known to stalk medical practice staff if they are dissatisfied with the treatment or diagnosis they have received. Patients more often assign blame to the nearest person at hand, either physician or staff. This activity should be taken seriously and appropriate authorities notified.

DOMESTIC VIOLENCE

Because of the rising incidence of domestic violence, one of the greatest threats to employee safety can come from their own homes when domestic violence follows them to work. Women who suffer from domestic abuse carry the burden of their situation into work every day. Domestic violence impacts the workplace because the one place an abuser knows he can find his victim is at work.[11]

In 1997, the US Department of Justice published a law school report on educating to end domestic violence.[12] The report provides comprehensive information, facts, and preventive measures involving domestic violence.

Violence is a societal problem of epidemic proportions. Experts estimate that 2 to 4 million American women are battered every year, and that between 3.3 and 10 million children witness violence in their homes. Battering affects families across America in all socioeconomic, racial, and ethnic groups. As information about the extent and impact of domestic violence emerges, it has been identified as a criminal justice issue, a public health crisis, and a costly drain on economic productivity.

Domestic violence is a pattern of behavior that one intimate partner or spouse exerts over another as a means of control. Domestic violence may include physical violence, coercion, threats, intimidation, isolation, and emotional, sexual, or economic abuse. Perpetrators often invent complex rules about what victims or the children can or cannot do and force victims to abide by these frequently changing rules.

Domestic violence is not defined solely by specific physical acts, but by a combination of psychological, social, and familiar factors. In some families, perpetrators of domestic violence may routinely beat their spouses until they require medical attention. In other families, the physical violence may have occurred in the past; perpetrators may currently exert power and control over their partners simply by looking at them a certain way or reminding them of prior episodes. In still other families, the violence may be sporadic, but may have the effect of controlling the abused partner.

Studies have shown that the only common traits between victims are that they are being abused by their intimate partners or spouses and that the majority of heterosexual victims are female. Victims may be doctors, business professionals, or nurses, among others. Perpetrators may be police officers, sports heroes, CEOs, lawyers, or college professors, among others. Unlike victims, perpetrators do have at least two common traits—the majority of perpetrators witnessed domestic violence in their families and are male.

Victims or perpetrators cannot be identified by a certain profile. Abusers may appear charming and articulate when confronted or they may be seething with rage; similarly, victims are just as likely to seem angry or aggressive as frightened or passive.

Employers can help provide safety at work by implementing a domestic abuse program that includes policies and training for supervisors. Preventive measures include:

- The employer should obtain a photograph of the abuser so the batterer will be recognized if he or she shows up at the workplace. He or she should be asked and required to leave or a law enforcement agency should be called.

- The employee should keep a copy of a protection order at work. The employer should have a copy of the order.
- Screen calls to the victim with voice-mail or a machine, if possible. An attorney may be able to introduce recorded threats made by the batterer as evidence in a court case.
- The victim should travel with another person. Victims frequently are harassed on the way to or from work by batterers who are jealous of coworkers or want victims to lose their jobs and become economically dependent on them.

Domestic violence in the workplace provides a significant challenge to employers who seek to protect employees and to limit their liability. The challenge is to extend the net of protection to include all foreseeable threats to the workforce. Foreseeability, in the law, may equal liability. Liability attaches when a responsible party either knew or should have known of a particular hazard and thereafter failed to address it.[13]

RECOGNIZING RISK FACTORS

Newspapers include stories about disgruntled workers who feel the only recourse to a wrong is to violently respond, often leading to serious injury or the death of several colleagues. While studying violent people, researchers have developed psychological profiles of individuals who commit violent acts as well as what environmental factors have a negative effect on these individuals. By identifying both potentially violent people and situations in the early stages and correcting those factors that can be mitigated, both the number and intensity of violent incidents can be reduced. This section explores those environmental and human characteristics seen most often in violent situations and presents ways to lessen the risk of workplace violence to employees and patients.

Environmental Factors

As our society becomes more violent, workplaces mirror society at large. When establishing a workplace violence prevention program, management must give consideration to the practice's location in the community. If the clinic is located in an area where there are high incidences of drug-related crimes, thefts, burglaries, and other crimes, there is more of a chance that such activities will affect the staff. They may be injured while on the job when there is another crime being committed. With the large number of weapons in circulation in our society and the ease with which a weapon can be obtained, crimes often include weapons. Part of any workplace violence prevention program must assess the risk of violence occurring inside the medical practice in relation to the violence of its location. If there is a high risk, the staff should receive training in their options if they are faced with criminal activity.

Crime and violence are particularly apparent in neighborhoods or communities where there is gang warfare over supremacy—either for "territory" or the status of being "top dog." Access to drugs and handguns increases the risk of physical injuries and the stress of feeling unsafe at the workplace.

Should the practice—similar to banks or high-risk retailers—have systems installed with a silent alarm? Would a security guard service be more practical? What other options might there be? These issues should be discussed with the staff. Employee training should also be conducted to enforce sound safety procedures.

Another factor to assess is the general physical layout of the medical practice. This includes:

- Ease of entry through doors and windows
- Alarm systems
- Lighting
- Soundproofing between exam rooms
- Intercom systems
- Appropriate separations between patients and staff
- Security of medicines, prescription pads, and proprietary items
- Other safety measures

If employees are lax about locking a rear entry, convert it to a self-locking door. If the waiting room is difficult to monitor, consider convex mirrors.

In many communities today—even affluent ones—keeping hallway doors and outside restrooms locked is commonplace. However, a lock is insufficient protection if an employee is careless. For example, allowing a door to close on its own is an invitation for a criminal to gain access before the door locks. Likewise, leaving the bathroom key on a ring that also holds keys to other areas allows the criminal easy entry to those other areas. In many instances, leaving the key on a countertop or table makes it easy to steal. And it is not just thieves who favor the hiding benefits of a restroom, it is also a popular spot for rapists.

Enlist the aid of staff members to isolate problem areas and to come up with solutions to improve staff and patient safety. Is there a need for brighter parking lot lights? Should a buddy system be instituted for employees working after dark? What other potential or real problems need to be addressed for a viable workplace violence prevention program? Are hedges close to parking spaces? Do they need to be trimmed?

Another safety precaution is to network with other businesses in the area of the practice. Coordinating efforts to minimize crime, as well as seeking the advice from the police and other community support organizations in the area, are proactive approaches to reducing the risk of societal violence.

By discussing problems with other business people in the area, the medical practice can stay abreast of current crime trends or acts of violence. This information should be shared with the staff members so they are alert to the possibilities. The employees must also know when to contact law enforcement agencies and what to do to help. All staff members should be regularly updated as a part of any workplace violence prevention program.

Organizational Factors

Research indicates that violence between staff members (eg, threats, harassment) occurs more often in organizations with weak or nonexistent human resource practices and policies. This is particularly true where the company has an inadequate policy regarding workplace violence prevention or does little or nothing to enforce existing policies.

Sometimes there are no clear rules of business conduct in organizations, and employees often push behavior to or beyond the limit of what is acceptable. Medical groups should implement procedures for reporting employees with problem-behavior. The procedure should maintain worker confidentiality, allow for immediate review, and outline possible actions that will be enforced. The procedure's purpose is to improve staff safety and deter unacceptable behavior. Once management establishes the rules and how they will be enforced, they must convey them to all staff members. A signature sheet should be on file indicating that each employee read and understood the information. Failure to enforce the rules is a license to break them.

If the policies have not been consistently enforced, this practice must change. Policies must be applied equally throughout the organization, regardless of a worker's position within the clinic. Spotty enforcement can create a hostile work environment, increased staff stress and turnover, decreased productivity, increased employer costs of hiring and training, and other employee relations problems. When staff members feel they are not respected, they are more likely to react in antagonistic ways, including sabotaging, stealing work-site property, or selling or sharing practice information to a competitor.

Establishing policies and guidelines that define acceptable behavior within the work site, training all employees in these practices, and proactively enforcing them increases employee safety. Employee productivity and loyalty increase when they feel safe and believe their employer and coworkers value them. The following are basic needs of people, whether at work or at home[14]:

- To know they are valued
- To be told the right answer
- To be treated as if their requests were reasonable
- To be treated fairly and equitably
- To be treated as individuals
- To be treated in a friendly, courteous manner
- To be kept informed
- To be informed in an understandable manner
- To be recognized, feel important, and be acknowledged
- To have complaints listened to and adjustments made
- To be thanked

Human Factors

People have virtues and strengths as well as faults and weaknesses. Knowing some of the indicators of a potentially violent individual can reduce the chances of an act of aggression.

In practices where there are many employees, patients, visitors, vendors, and delivery personnel, it is often difficult to keep track of everyone who enters or leaves. It is incumbent upon the staff and the patients, for their own well-being, to be vigilant.

It is not unusual for an upset patient to snarl at the receptionist but appear to be charming and affable to the medical staff. The medical staff members would have no way of knowing that this patient is disguising hostility unless the receptionist alerts them. Once inside the examination room, anger might escalate.

It is possible to develop an awareness of what might push a person to violence and take action to reduce the possibility. One of the best ways and least "obvious" is to listen to conversations between people and be observant, whether on staff or not. This is not a matter of eavesdropping so much as it is monitoring a tone of voice, an aggressive manner, or even a threatening movement. It is not to be assumed that all human rage is rooted in a loud voice; people evidence anger in different ways. The lion may roar, but the cobra hisses.

Clenched fists, a forward thrust of the head, or pointing a finger can be forewarnings of anger that can lead to violence. If possible, the medical staff should be apprised of the nature of the patients' illnesses, what medications they are on, and whether or not the patients have been under undue stress. If the staff members are aware of the patients' problems and take note of any hostile attitudes, they are better able to anticipate possible acts of violence and handle them before they get out of control.

Some personality traits appear to be linked to violent behavior more than others. For example, people who never accept responsibility for their own actions and constantly assign blame to some other cause or person tend to be more violent. They often express contempt for bosses or other authority figures. They have a tendency to rationalize their behavior as if the other person "had it coming." These people may also intimidate coworkers.

Other characteristics include employees who feel they are entitled to special privileges or continually compare what tasks they are given in relation to others on staff. The media has made much of people who are "loners" as potentially violent. This may lend itself to misinterpretation; there is a difference between "quiet" people and persons considered to be "loners." Loners probably never have lunch with anyone, might nod their heads in acknowledgment—but only if the other person says hello first, do not speak to coworkers about anything other than work, never talk about their personal lives, and so forth. Quiet people, however, speak when they have something to say, but aren't conversationalists by nature. They lunch with coworkers, more often listening than speaking. If something exciting happens in their lives, they openly share it. They are unlikely to let a door slam shut or drop something instead of placing it. They are . . . quiet.

A history of untreated or unresolved emotional or psychological problems or a need to seek revenge are factors for violent behavior. Employees or patients with a substance abuse problem, including alcoholism, are more likely to react negatively under stress.

Remember that stress is a *perceived* response. Sometimes it is justified, and sometimes it is not. Levels of stress are how an individual responds to a situation that is viewed as unwarranted, threatening, or in some manner detrimental to the individual's well-being. What constitutes stress for one person could be just a fact of life to someone else.

People who resort to violence to solve problems often have obsessive and/or compulsive behaviors that may involve weapons, paramilitary training, or romantic infatuations. This is especially seen when domestic violence intrudes upon the workplace. When people bring obsessions or personal stress factors into the workplace because they are unable to set them aside, it can indicate more serious problems.

Symptoms of depression that are not related to a depressive illness or event are often red flags or signals. Behavior such as an inability to concentrate, sudden downswings in job performance or attendance, poor behavioral control, or inattention to personal hygiene could be warning signs. Additional signals might be sudden withdrawal from friends, family, and peers and may be indicators that a need for intervention exists, especially if the behavior continues over a period of time.

Keys to reducing violence include establishing a system to deal with the problem and preparing contingency plans. Case management conferences with employees help increase awareness of potential problems and can provide staff with countermeasures to prevent or reduce aggression. Staff training should include violence and aggression prediction, personal protective measures, anger management, using specialized responders when the risk of violence is high, and how to deal with hostile individuals other than patients.

ORGANIZATIONAL RECOVERY AFTER AN INCIDENT

Whether the incident is a newsworthy event or threatening behavior between two staff people contained within a small work group, there is a need to help employees return to the workplace routines. There are several ways in which management and supervisors can help the staff return to normal routines. Depending upon the incident's severity, management should take appropriate intervention by being present and sharing as much information as possible with employees. Confidentiality may become an issue sometimes, yet management awareness and involvement can help stem rumors and prevent more problems.

Whenever possible and applicable, include union representation in all discussions and interventions. When the incident warrants extra steps, bring in crisis-response professionals to help support staff and to screen for people who might require more intensive counseling to cope with the trauma.

Witnessing an act of violence involving injuries or death and coping with the scene or the memory can be extremely disturbing and upsetting. It is vital that management actively supports the debriefing process used by the crisis-response team. Management may need to buffer those most affected from postincident job stressors by reducing workload or length of the workday for a period of time. Caregiving in this way shows that staff members are respected and that management is sympathetic to their needs. This helps medical personnel to more quickly put the incident in perspective and return to a normal routine.

Additionally, there may be staff members who have difficulty returning to the site of a violent incident and may need more intensive help. No two people respond to trauma in the same way; people heal in their own time. Employees who have experienced the violence either as the target or the witness may have delayed reactions. Work can help the employees refocus and begin the healing process, but it should be balanced with support and awareness of the impact of the crisis. For assistance, consult a mental health professional who is trained to intervene during the aftermath of workplace violence.

A WORKPLACE VIOLENCE PREVENTION PROGRAM

In 1996, nearly one million Americans were victims of workplace violence at a cost of more than $55 million in lost wages.[15] Developing a sound workplace violence prevention program makes sense from both a cost-benefit and worker safety perspective. It is also the right thing to do.

Six-Step Workplace Violence Prevention Program

The following is a composite of several of the many workplace violence prevention programs that are available to help employers:

1. *Select a crisis management and prevention team.* Include a representative from all facets of the group's practice: doctor, nurse, technician, management, clerical staff, union representative (if applicable), and perhaps a patient or patient advocate. They should regularly meet in order to review and make recommendations for annual or semi-annual changes.
2. *Conduct a violence vulnerability audit within the workplace and the neighborhood and brief senior management on its results.* Evaluate all potential hazards and violence risks as to their level and intensity of occurrence. These risks include entering and leaving the office, the likelihood of a patient losing control, an incident between staff members erupting into violence, the current environment within the facility, the type of management style, and the daily job stressors. Also, during the audit, meet and establish a rapport with local police and other businesses. Once complete, report these findings to management.
3. *Establish an antiviolence policy and train the staff on it.* Use the audit to create a workplace violence prevention policy that includes acceptable codes of conduct and speech, reporting and confidentiality procedures, disciplinary options, enforcement protocol, and guidelines to follow after an incident that involves a major trauma situation. Communicate this policy to the staff through training in such areas as anger management. Incorporate annual refresher courses or training workshops that pinpoint various issues as they develop. Role-playing, with employees taking turns as outraged or violent patients, may also be a valuable training tool. Such role-playing can help employees learn how to handle situations, for example, what to do if faced with a bombing, kidnapping, or shooting; who is responsible for media coordination or contacting victim's families; and what is the response to a worker's death or injury.
4. *Immediately investigate and correct incidents.* Failing to take action after a problem is identified hurts staff morale. The perception of unfairness may even trigger aggressive behavior, creating a more widespread problem. Clearly and promptly carry out disciplinary

measures so the situation does not escalate. According to the 1970 Occupational Safety and Health Act, employers have a responsibility to ensure a safe work environment. If an employer or supervisor is aware of a potentially violent or unsafe situation and does nothing to intervene, the employer may be found liable for any injuries or deaths that result. Keep records regarding incidents and the corrective actions that were taken.

5. *Retain a trauma response team that can be immediately activated.* Large medical practices and private clinics may want to consider retaining a trauma (or crisis) response team to come into the workplace and counsel those affected in a violent situation. Such counseling can reduce employee stress. Although some medical practices have an employee assistance program (EAP) for more short-term counseling and referral, trauma counseling requires professional mental-health specialists who are specially trained in helping people cope with a traumatic situation.

6. *Employers of high-risk practices should consider additional safety precautions.* After audits are completed and procedures developed, it may be clear that the practice is a potential target or at high risk for a workplace violence incident to occur. Should this be the assessment, additional safety precautions beyond common sense and specialized training may be necessary. Consultations with safety or security specialists can help in the development and implementation of auxiliary precautions.

OSHA'S INVOLVEMENT

The General Duty Clause of the 1970 Occupational Safety and Health Act requires employers to provide a safe and healthful working environment for all workers covered by the act. At that time, the concept of "safe and healthful" was most often associated with issues of toxic substances or dangerous physical tasks.

In addition to compliance with hazard-specific standards, all employers have a general duty to provide their employees with a workplace that is free from recognized hazards likely to cause death or serious physical harm. Consequently, OSHA also relies on Section 5(a) of the General Duty Clause for enforcement authority for workplace violence incidents. Employers may be cited for violating this clause if there is a recognized hazard of possible workplace violence in their practice and nothing is done to prevent or diminish that threat. Employers may also face litigation by employees or their families who are the victims of workplace violence.

Escalation in workplace violence in all fields, not just medical practices, increases the need to understand its causes and examine the possible preventive measures. This led OSHA to focus on workplace violence, especially those violence-related concerns that increase the stress levels of staff members.

In 1998, OSHA published *Guidelines for Preventing Workplace Violence for Health Care and Social Service Workers*[16] following a study that showed health care workers were at high risk for experiencing a workplace incident of violence. The guidelines include OSHA's four main components to any effective safety and health program, which also apply to preventing workplace violence. The program elements are:

- *Management commitment and employee involvement.* Working together are essential elements of a program.
- *Workplace analysis.* A common sense analysis of the workplace to find existing or potential hazards for violence is essential.
- *Hazard prevention and control.* This step designs measures through engineering or administrative and work practices to prevent or control the identified hazards.
- *Safety and health training.* Ensuring that all staff are aware of potential security hazards and how to protect themselves and their coworkers through established policies and procedures is the crux of this step.

Workplace violence can no longer be viewed as a random act. Virtually no occupation is free from the possibility that a staff member might be the victim of workplace violence. Advance preparation and knowledge may make the difference between an argument or a murder. Adopting practical measures can significantly reduce the serious threat to employee safety and can provide employees with skills to help them choose to step back from an angry confrontation or remain calm in a threatening situation. Management can help establish an atmosphere of trust and knowledge that leads employees to make the right life-saving decisions.

ENDNOTES

1. Occupational Safety and Health Administration. US Department of Labor. *Guidelines for Preventing Workplace Violence for Health Care and Social Service Workers.* OSHA 3148, 1998.
2. Bachman R. US Department of Justice, Office of Justice Programs, Bureau of Justice Statistics. Violence and theft in the workplace. *National Crime Victimization Survey.* July, 1994. NCJ-148199.
3. Bureau of Labor Statistics. *Fatal Occupational Injuries Resulting from Transportation Incidents and Homicides by Industry* (1997 and 1998). Washington, DC: US Department of Labor.
4. Warchol G. Workplace violence. *1992-96. Bureau of Justice Statistics, Special Report.* July, 1998.
5. National Institute for Occupational Safety and Health. Violence in the workplace: risk factors and prevention strategies. *Current Intelligence Bulletin 57.* June, 1996.
6. Jackman J, Onyango C, Gaurilles E. *1998 National Clinic Violence Survey Report.* Feminist Majority Foundation. Released January, 1999.
7. *Thapar v. Zezulka*, No. 97-1208. 994 S.W.2d 635.
8. *Tarasoff v. Regents of the University of California.* 551P22d 334 (1976).
9. Tjaden P, Thoennes N. *Stalking in America: Findings From the National Violence Against Women Survey.* Research in Brief. Washington, DC: US Department of Justice, National Institute of Justice; April, 1998. NCJ 169592.
10. Orion D. *Stalkers & Stalking* in: *I Know You Really Love Me: A Psychiatrist's Journal of Erotomania, Stalking, and Obsessive Love.* New York: Macmillan; 1997.

11. Boyle J. Domestic violence: from behind closed doors to the company floor. *Today's Supervisor*. National Safety Council. Itasca, IL; August, 1999.
12. Goelman D, Valente R. *When Will They Ever Learn? Educating to End Domestic Violence*. A Law School Report. Chicago: American Bar Association; 1997.
13. Chavez LJ. *Domestic Violence in the Workplace . . . A Puzzling Issue for Employers*. 1997; http://members.aol.com/endwpv/Pkg-Deal.html.
14. Chaff LF. *People Power: A Customer Service Program for Employees in Healthcare and Other Industries*. Chattanooga, Tenn: Chaff & Co; 1992.
15. National Institute for Occupational Safety and Health. Violence in the workplace: risk factors and prevention strategies. *Current Intelligence Bulletin 57*. June, 1996.
16. Occupational Safety and Health Administration, US Department of Labor. *Guidelines for Preventing Workplace Violence for Health Care and Social Service Workers*. OSHA 3148, 1998.

chapter 12

Indoor Air Quality

Twinkle, twinkle, little bat!
How I wonder what you're at!
Up above the world you fly,
Like a tea tray in the sky.
Twinkle, twinkle.

Sang by the Mad Hatter at a great concert given by the Queen of Hearts[1]

The Mad Hatter in *Alice in Wonderland* appeared to be a whimsical character, nothing more; however, the Mad Hatter's eccentric behavior could have resulted from his use of mercury while creating felt hats for gentlemen. The theory offered is that by using mercury on a daily basis, he was imbalanced from the prolonged exposure to the fumes of his trade. The fumes were an occupational hazard.

Likewise, house painters have long known that paint fumes affected them and believed drinking beer could help offset the ill effects. Coal miners know the consequences of working deep within the mines. Black lung disease was also considered just an occupational hazard.

These circumstances, except for the mines, existed when windows could be opened whether at home or at work. "Airing the place out" was often considered an adequate means of dealing with pungent odors caused by chemicals. However, the problem was viewed as the smells and not their possible effects.

OVERVIEW OF INDOOR AIR QUALITY

Indoor air pollution poses many challenges to medical practices. Air pollution, both outdoor and indoor, is linked to a number of serious health issues including inflammation of the respiratory tract, decreased lung function, premature aging of the lung, and increased morbidity and mortality in patients with lung or heart disease. Breathing problems caused by the workplace can range from a stuffy nose, laryngitis, asthma to emphysema and lung cancer.

Prior to the EPA's existence, attitudes toward indoor air quality (IAQ) were primarily confined to chemicals in enclosed industrial workplaces. Unlike the industrial environment in which a chemical exposure can cause a consistent set of complaints and symptoms among a relatively large group of workers, poor IAQ is often noted by isolated and seemingly unrelated complaints. For example, an individual may complain about an odor that no one else notices; other people may complain that they have difficulty breathing; while still others may claim that each time they enter the building, they have a headache. Some people are allergic to certain fabrics and may suffer effects from their cubicle coverings or even the carpeting in their office.

In the past, such individual negative complaints were overlooked or dismissed as nonsense. Yet, as they gained more experience and knowledge, researchers and building owners began to understand that the isolated complaints might be related to IAQ.

In an effort to conserve energy, modern building design favored tighter structures with lower rates of ventilation (ie, recirculating air for heating, cooling, ventilation). Building construction ensured that little or no outside air came into the building.[2] While this practice has since been dismissed—primarily because of IAQ concerns—there are still structures in the United States that rely on such technology for heating, cooling, and ventilation. Sealing off a structure to almost all outdoor air may be more cost- and energy-efficient, but the trade-off is increased airborne pollutants that often lead to employee health problems (ie, headache, allergic skin reaction, fatigue, nausea, persistent cough, neurological dysfunction).[3]

Advances in building design, construction, and maintenance have made it easier for owners and tenants to redesign their spaces to suit their particular needs. This flexibility often includes installing windows, removing old wallpaper, painting, recarpeting, and tiling. These activities may contribute to poor IAQ and health complaints, which may appear to be vague or even groundless. However, they are very likely rooted in the workplace air quality.

As it became more apparent that the general population has a wide range of susceptibility to the effects of biological allergens and chemical pollutants, silent accusations of probable hypochondria gave way to a new realization. Not everyone physically reacts to everything in the same way. NIOSH, EPA, CDC, the World Health Organization (WHO), and other agencies and organizations have begun more aggressively funding research agendas to study indoor environmental quality to protect public health from the harmful effects of indoor air pollution and to eliminate or reduce contaminants that are known to be, or likely to be, hazardous to human health and well-being.

TYPES OF COMPLAINTS

EPA and NIOSH have identified the more common complaints related to IAQ:

- Headache
- Fatigue
- Chest tightness
- Shortness of breath
- Sinus congestion
- Coughing
- Sneezing

- Eye, nose, and throat irritation
- Skin irritation
- Nausea
- Dizziness

Some of these complaints may be related to allergies, while others are more often associated with chemical exposures. The most vexing problem is when there does not appear to be an easily identified, clearly discernible source. If employees perform tasks not usually part of their duties and complain about symptoms associated with IAQ, isolating the cause may be somewhat easier, but confines the problem to an occasional complaint from one or two employees.

If there are serious illnesses or adverse health effects among coworkers, the problems may be due to a more serious situation. When several employees have the same serious complaints, problems become more difficult to deal with because of the emotional effect they impart. If staff members report problems such as miscarriages, reproductive problems, or birth defects, the local or state health department should be consulted. While any complaint requires an inquiry into the probable cause, multiple complaints may well be a warning that there is something seriously harmful to the health of the employees. An industrial hygiene and epidemiological investigation might be required. Any work-related health threat that is likely to have permanent effects, either physical or emotional, requires immediate attention and corrective action.

POTENTIAL SOURCES OF BUILDING-RELATED ILLNESSES

Two components may be responsible for health complaints in a medical office: the building and work-related tasks.

The first component is a problem with the building itself. Old insulation, rusty pipes, lead-based paints, dirty air filters, concrete-block structures, and so forth often result in allergenic reactions. If the practice is located on the first floor that sits on slab concrete, it can hold in dampness and pollutants and absorb odors or elements of anything that is spilled. The interior and exterior of the building itself cannot be ruled out as a possible cause.

The other component is the effects of work-related tasks. There may be activities in or around the building producing pollutants that adversely affect the staff. For example, if the ground maintenance crew mows or prunes, airborne pollutants can find their way inside the building. Construction or remodeling that is taking place on another floor or next door can send dust and other particles into the ventilation system.

Numerous tasks and circumstances within a medical practice can lead to unhealthy working conditions. For example, if the practice includes a lab, technicians use chemicals. Indoor air pollution may include the use of chemical sterilants such as gluteraldehyde or formaldehyde; anesthetic gases, such as nitrous oxide; latex gloves that can pose a problem for people with hypersensitive skin; organic solvents such as xylene (used in histology laboratories); and even a coworker's aftershave or cologne, which can be a health problem in some instances.

IDENTIFYING BUILDING-RELATED PROBLEMS

A significant source of health complaints relates to the HVAC systems in a building. Modern HVAC systems are intended to efficiently heat, cool, and add fresh air to a building while simultaneously removing pollutants. However, not all structures have modern HVAC systems. Figure 12.1 shows a typical HVAC system.

Geographical location can also play a role in the HVAC's efficiency. In some states, very high humidity levels only exist for a few weeks, while in other states, those levels can be present for months. Similarly, freezing temperatures can put a strain on any HVAC system. For example, the type of HVAC needed for a structure in Minnesota may not be necessary for a similar building in Louisiana. Air quality control in New York City probably is considerably different from that in Vail, Colorado.

In addition to regional weather differences, certain structures benefit the HVAC system more readily than others. Wooden structures (eg, clapboard) do not retain heat or cold as well as brick or stone; therefore, these buildings probably require a customized HVAC system to accommodate their structure. Depending on the building's size, construction, and age, an HVAC system is likely to have numerous components, including:

- Heat pumps
- Furnaces
- Boilers
- Chillers
- Cooling towers
- Steam lines
- Water lines
- Air-handling units
- Duct work
- Exhaust fans
- Filters

A major reason for the efficiency of modern HVAC systems is that they recirculate a significant amount of indoor air. That recirculation allows the system to use less energy to heat or cool the building. That very efficiency, however, can cause the following problems:

- Areas of excessively high temperature
- Areas of excessively low temperature
- Low humidity
- High humidity
- Concentration and recirculation of pollutants
- Circulation of airborne bacteria

Mold growth due to improper water drainage or entrapment in ventilation systems is a major contributor to poor IAQ and occupant illness. In older buildings, heating and ventilation ducts are often lined with fiberglass insulation to increase heating and cooling efficiency. However, fiberglass can trap moisture and dirt particles, creating an ideal growth host for mold and bacteria.

FIGURE 12.1

A Typical HVAC System Used in Modern Buildings

Reproduced from US EPA Indoor Air Quality: Tools for Schools, 1995.

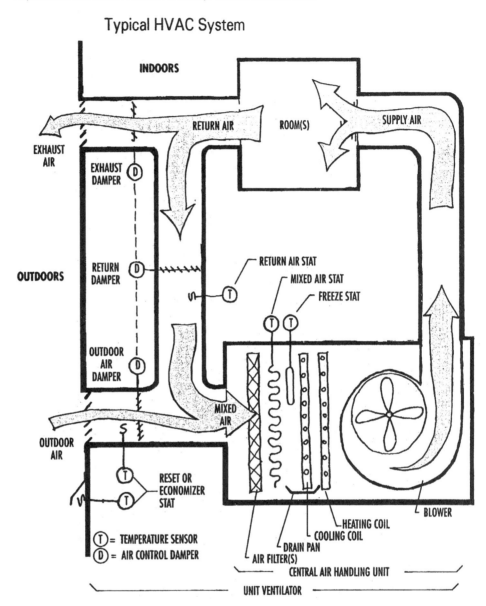

Properly maintaining an HVAC system includes regular cleaning, particularly drains and drip pans. Standing water in drain pans and other parts of the HVAC system can be the source of Legionnaires' disease.

External Activities

Though modern buildings do recirculate a considerable amount of air, they also bring in fresh air. Importing fresh air from outside is a problem when air intakes are located near parking garages, heavily trafficked streets, or other areas that are laden with exhausts that contaminate the air. As a result, when the so-called fresh air enters a building, exhaust fumes or industrial pollutants may enter as well. This is also true of outdoor smoking areas. If air intakes are located at or near outdoor concentrations of pollutants, the intakes cease to have the desired effect.

Medical practices are often located near freeways, busy thoroughfares, under airline flight paths, or near railroad tracks. Transportation methods (ie, cars, trucks, airplanes, trains) increase the amount of pollutants in the air, often to very high concentrations. On hot, humid days, the dangers from these pollutants measurably increase and can prove fatal.

Another common problem, especially with multiuse buildings, is the proximity of businesses that rely on high concentrations of chemicals in the course of their functions, such as dry cleaning establishments, nail salons, most industrial mills, public restrooms in transportation stations (eg, bus, train), photography labs, graphic arts studios, and newspaper or other types of printing press facilities. Poorly designed ventilation systems can bring chemical fumes and odors from those businesses into medical practices as part of the "fresh air" delivery system. So while the health problems may be happening within the building or medical practices, the cause may well be from next door.

When establishing or relocating a medical practice, look beyond the confines of the floor space. Prior to signing a lease, check the following:

- Location of air intakes to avoid proximity to parking garages, busy intersections, or transportation methods
- Sources of fresh air (intakes) that are not near vented or exhaust air from businesses that use chemicals as a routine aspect of conducting business
- Designated smoking areas are located well away from air intakes

Also rely on instincts and judgment. The following questions are areas of consideration during a walk-through of a potential new location:

- Does the temperature feel comfortable?
- Does the lighting seem appropriate?
- Do noise levels seem reasonable?
- Are there odors?
- Does the building appear clean?
- Are there any disconcerting vibrations?
- Do eyes feel dry?
- Is there noticeable nasal, skin, or respiratory irritation?

Internal Activities

There are many internal activities in a medical practice that may cause poor air quality.

Office Décor

Décor may be a significant contributor to poor IAQ. Dust accumulates on all surfaces but is a particular problem when it lands on draperies, blinds, and carpeting. A vacuum with a high-efficiency particulate air (HEPA) filter can mitigate this problem; however, consideration should be given to office space that is free from carpeting and heavy draperies.

Carpeting can be the source of mold if laid directly on a cold surface, such as unventilated concrete that causes moisture to condense under the carpet. Vinyl wallpapers may also be a source of mold growth because the paste that applies it to the walls can trap moisture between the paper and the wall.

Environmental Tobacco Smoke (ETS)

In the absence of sufficient ventilation, chemicals from building materials and stored chemical products, such as bacteria, dust, and insects, can contaminate indoor air. One of the most harmful and widespread contaminant of indoor air is tobacco smoke. Public concern about environmental tobacco smoke has grown considerably in the last decade. Communities continue to develop new laws and policies that ban or restrict smoking in many workplaces and public buildings. ETS is a combination of exhaled smoke and the smoke produced by an idling cigarette, cigar, or pipe. It consists of solid particles, liquids, and gases.

Scientists have identified more than 4,000 different chemical compounds in ETS, including nicotine, carbon monoxide, ammonia, formaldehyde, and dioxins. More than 50 of these substances are known carcinogens. Others are known or suspected mutagens capable of changing the genetic structure of cells. The EPA has recently declared ETS to be a Class-A carcinogen.

Exposure to ETS for brief periods can produce eye, nose, and throat irritation; headaches; dizziness; nausea; coughing; and wheezing. ETS can markedly aggravate symptoms in people with allergies or asthma. Long-term exposure has been linked to heart disease and cancer. In young children, exposure to ETS can result in chronic respiratory illness, impaired lung function, and middle ear infections. ETS can retard the growth and development of fetuses, resulting in low birth weight and a greater likelihood of complications during pregnancy and delivery.

Pest Control

Pest control methods often depend on the use of pesticides, whose storage, application, and handling can have serious health effects if label instructions are not followed.

Remodeling and Renovation

Remodeling and renovation activities are a significant source of indoor air pollution. Dust from demolition and dry-wall finishing, paint fumes, carpet, sanded tile glues, and fabric finishes on new draperies and upholstery can be drawn into a ventilation system and diffused throughout the building. Along with airborne dust, odors can also seep into the building's air supply. While perhaps not a direct health threat, odor can be mildly to downright unpleasant. Decaying rodents that get caught in the wall spaces of old buildings can also create a pungent odor.

Relocating a Practice

If locating a practice in a building more than 25 years old, determine if it was constructed using asbestos products for pipe insulation, glues under floor tiles, or lead-based paint. Older structures may have the following possible concerns: fuel-oil furnaces can create noxious fumes; water pipes may have rusted, contaminating drinking water; certain types of foam insulation can prove unhealthy; and so forth. These types of situations require special consideration and attention from safety and health specialists.

If the medical practice uses chemicals, such as gluteraldehyde, anesthetic gases, or organic solvents in histology laboratories, talk with building management to ensure that those substances will be exhausted directly to the outside.

APPROACHES FOR SOLVING INDOOR AIR QUALITY PROBLEMS

General solutions to IAQ problems can often be determined with basic investigation in and around the medical practice building. These basic investigative steps include:

- Making sure that all vents (ie, intake and outtake) are free from obstructions, such as furniture, and are clean
- Checking that all filters are clean
- Keeping grill covers free from dust and debris
- Routinely checking ducting for dust or other factors that may affect air quality

Routine Housekeeping and Maintenance

Ordinary housekeeping can cause settled dust to become airborne and introduce chemical fumes from cleaning products. Table 12.1 offers housekeeping steps to use in order to reduce IAQ problems. Some vacuum cleaners rely on suction intake systems while others depend on a combination of intake and exhaust. The latter will pick up particles, but emit fumes back into the area.

If housekeeping activities are suspected as a source of poor IAQ, consider the following:

- Review the cleaning products in use and try to find odorless substitutes.
- Keep a special closet or cupboard in a separate, well-ventilated area to house all cleaning products.
- Use vacuum cleaners equipped with HEPA filters.
- If possible, open all windows and doors if no obvious pollution-producing activities are occurring outdoors.

TABLE 12.1

Specific Housekeeping Steps Used to Reduce IAQ Problems

- Prepare and follow written housekeeping procedures that detail the proper use, storage, and purchase of cleaning materials.
- Be aware of the housekeeping products and equipment used in your building, particularly those that are potential irritants or have other IAQ effects.
- Purchase the safest available housekeeping products that meet your cleaning needs.
- Educate housekeeping staff or contractors about proper use of cleaning materials, cleaning schedules, purchasing, materials storage, and trash disposal.

Reproduced from US EPA & NIOSH, *Building Air Quality Action Plan*, June 1998.

Routine maintenance may involve painting or chemical use for short periods of time. If those activities must be performed during office hours, make sure the workers take the following steps:

- Use local exhaust ventilation to remove fumes and dust
- Close air intakes before working with paints or chemicals outside the building

Pest Control

Chemical pesticides must be dealt with carefully to avoid IAQ problems. One way to minimize the risk of IAQ problems from pest control is integrated pest management (IPM), which emphasizes the use of non-chemical pest management practices wherever practical. Table 12.2 describes the IPM process. If possible, chemical pesticides should be applied during nonworking hours to reduce IAQ problems. For example, if the medical practice is closed weekends, applications on Friday nights provide increased ventilation over the weekend.

Waste Management

Proper waste management promotes good IAQ by controlling odors and reducing contaminants. It is also an IPM method that controls vermin. Furthermore, good sanitation practices decrease the need for pesticides.

When selecting the proper waste containers, consider the kind of waste that will be placed in them:

- Food waste or food-contaminated papers and plastics should be contained securely (eg, covered containers, tied-off plastic bags) to discourage flies and other vermin.
- Recycling and other special handling bins should be clearly labeled so other types of trash are not put in them.

Locate dumpsters away from outdoor air intakes, doors, and operable windows. Regularly empty waste containers and store them in an appropriate location. Follow a regular schedule that minimizes odors and deprives vermin of their food source. The following are suggestions for routine waste management tasks:

- Containers with plastic liners that are regularly replaced do not usually need to be cleaned and disinfected as often as unlined containers.
- Do not store waste containers in rooms that have heating, cooling, or ventilation equipment that supplies conditioned air to other rooms.

TABLE 12.2

Use Integrated Pest Management Whenever Practical

- Know what pest control products are used in your building.
- Prepare written pest contract procedures that detail the proper purchase, use, mixing, storage, and disposal of pesticides according to label directions.
- Use nonchemical pest control strategies where possible.
- Purchase the safest available pest control products that meet your needs.

Reproduced from US EPA & NIOSH, *Building Air Quality Action Plan*, June 1998.

Environmental Tobacco Smoke

While smoking is rarely a problem in medical offices, it is important to enforce a policy of nonsmoking. The effect of smoke particles in the gaseous components of environmental tobacco smoke (ETS) is unknown. Establishing a smoking policy that protects occupants and visitors from exposure to ETS is essential to maintaining good IAQ in the medical practice. Removing the health risks of ETS exposure requires a concerted effort to eliminate tobacco smoke from the workplace and other enclosed environments that the workers occupy. Examples of controls for reducing exposure to ETS include:

- Increase ventilation, which dilutes the smoke. However, this procedure does not make smoking safe because there is no known safe level of exposure to carcinogens.
- Restrict smokers to separate rooms, if the rooms have their own ventilation systems.
- Install electronic air filters and air purifiers, which may remove some smoke particles from the air. These controls cannot remove smoke that has settled on food, furnishings, skin, and other surfaces.
- Remove the source by instituting a policy that prohibits smoking in the workplace.

Renovation and Remodeling

Renovation and remodeling activities are significant sources of indoor air pollution. Demolition of walls generates dust. Removal of carpet also generates dust and may release molds. Painting generates fumes, as does installation of carpet. New carpet can also be a source of volatile organic chemicals. To avoid exposing the staff and patients to potentially unhealthy pollutants or to minimize the exposure, consider the following steps:

- Whenever possible, have all work done during nonoffice hours.
- During demolition, make sure the contractor uses appropriate dust suppression and removal methods, such as installing plastic sheet barriers to isolate construction areas, closing return air ducts in the work areas, and providing local exhaust ventilation to remove dust.
- During painting and gluing new carpeting, floor tile, or other floor coverings, increase the amount of fresh air entering the building. Ensure that the ventilation system operates even during nonoffice hours to help remove fumes during the drying process.
- When possible, store new carpeting, furnishings, and window treatments (eg, curtains, draperies) in well-ventilated storage areas prior to installation.
- Ventilate the remodeled area and the entire building before it is reoccupied, over a weekend or longer period, if possible.
- If reroofing or roof repairs occur during business hours, make sure the outdoor-air intakes are closed.

Clinical Practices

With regard to clinical practices and their need to improve IAQ, take the following basic precautions:

- Keep all sterilants and laboratory chemicals in closed containers in a suitable storage cabinet. The room in which workers use them should have a dedicated exhaust. Workers must be trained in the handling and use of all products.
- Ensure that workers who use the chemicals have the appropriate PPE and clothing for working with them. Provide periodic monitoring to be sure employees use required PPE.
- Routinely inspect delivering anesthetic gas apparatus for leaks.
- Provide substitutes for latex gloves, when necessary.

If the practice includes a laboratory that uses organic chemicals or equipment that requires chemical sterilization, those areas should have a dedicated exhaust system that does not allow for recirculation of those substances into the rest of the building. Filters that contain either activated carbon or zeolite (ie, a type of clay) absorb some volatile organic chemicals and should be considered. If using activated carbon or zeolite, follow the manufacturer's recommendations for replacement.

Multiuse Buildings

In a multiuse building, tenants have nominal control over IAQ other than setting the thermostat. This control also applies to maintenance work, repairs, or renovations. Good working relationships among the tenant, building management, and maintenance crews are essential. Maintenance crews are usually given work orders for the day, and a good rapport with management and maintenance crews can make the difference between having a ceiling tile replaced promptly or waiting until maintenance can get around to it.

Good working relationships with maintenance crews can also serve as safeguards against mix-ups, crossed signals, and other building maintenance problems that can occur when more than one person is needed to see a situation resolved, repaired, or routinely completed.

New Construction

When constructing a new building, work closely with the contractor and the person designing and installing the building's ventilation system. At a minimum, the ventilation system should follow ANSI/ASHRAE Standard Number 62-1989, *Ventilation for Acceptable Indoor Air*.[4] Discuss special needs with the contractor for dedicated ventilation to remove potential pollutants. The ASHRAE standards are not based on health. If the practice involves employees or patients with allergies or chronic respiratory problems, the addition of HEPA filters should be considered. Air intakes should be located so they do not bring vehicle exhausts or other fumes into the building. In addition, it may be prudent to have the site tested for radon gas and to ensure that the building design incorporates methods for the removal of this gas from the building.

MAINTAINING A HEALTHY INDOOR ENVIRONMENT

Diligence is an important aspect of maintaining a good IAQ program.

There are several ways the medical practice can help ensure a healthy indoor environment, and each of them requires some degree of involvement on the part of the practitioner and all of the staff. As previously described, if the practice is located in a multiuse building, a strong working relationship with the building owners and maintenance staff is important. Complaints and requests that are realistic, credible, and reasonable should be presented to the owners or building management. For example, if the parking lot is in the process of being repaved, there may be little that can be done about a smell associated with repaving.

One way to ensure the credibility of a complaint is to *listen* to comments from employees and patients about IAQ-related situations. Comments should be documented and include information such as the location of the complaint, time of day the complaint came in, and other relevant information. For example, if a certain odor occurs at 10:00 am but is not noticed throughout the rest of the day, timing can aid in narrowing down the source of the problem. Likewise, if there is a complaint about a specific area of the medical practice, but not throughout the premises, that, too, can help identify a problem source.

If written records of health complaints are kept, one helpful source to use to set up the records is the EPA/NIOSH action plan manual.[5] It has several forms that can be used to submit a complaint or to create an incident log, which can be helpful to the building's maintenance staff. A list of any chemicals used in the practice should be provided to the building's maintenance staff, along with a copy of the MSDSs for those substances.

Other safety and health approaches are as follows:

- *Identify and train a staff member as the health and safety coordinator.* It will be that staff member's responsibility to monitor IAQ complaints and symptoms, maintain a chemical inventory, track the effectiveness of housekeeping activities, and serve as liaison with building environmental management staff.
- *Hire an environmental consultant who specializes in IAQ.* The individual should be qualified to take measurements of temperature, humidity, carbon dioxide, air flow, and volatile organic chemicals. The individual should also be qualified to make recommendations for correcting IAQ problems.
- *Contact NIOSH, OSHA, or EPA to discuss the problem and seek resolution.* The medical practice can request that NIOSH conduct a health hazard evaluation (HHE). OSHA may also be contacted to test for air quality. If OSHA testing does not isolate the cause(s), the EPA should be contacted to investigate for outdoor pollutants.

Employees may present a series of nonuniform health complaints. The absence of uniform health complaints does not mean that the problems are individual. It simply means that people may react differently to the same environmental stressor. A good rapport among all employees at all levels is important so they understand that their complaints are being taken seriously.

Even though the HVAC system in the building may be sufficiently designed and properly maintained, occupant comfort may also be influenced by other factors that affect the system's operation. For example, the building may have been constructed using large glass panes for its exterior walls in order to reduce the need for interior lighting. When the sun shines on the glass, radiation heating may occur. Other factors that often influence occupants' perceptions of comfort include

- Poor lighting
- Excess noise
- Vibration
- Overcrowding of work areas
- Ergonomic problems
- Job stress

Though not considered a contributor to IAQ, excess noise and vibration can cause discomfort among building occupants. The sources of noise and vibration can range from heavy highway traffic on nearby roads to the air-handling equipment located on the roof of the building.

If air quality in the building is poor, employees may have a wide range of complaints that seem completely unrelated. Some of the complaints voiced by employees in office buildings may not actually be related to pollutants in the building. Solving IAQ problems requires a similar stepwise approach as diagnosing a patient's health problem. Table 12.3 suggests steps for responding to IAQ complaints.

TABLE 12.3

Establish Procedures for Responding to IAQ Complaints

1. Prepare and follow clear procedures for recording and responding to IAQ complaints, including:
    - Logging entries into your existing work-order system
    - Collecting information from the complainant
    - Ensuring the confidentiality of information and records obtained from complainants
    - Determining the response capability of in-house staff
    - Identifying appropriate outside sources of assistance
    - Applying remedial action
    - Providing feedback to the complainant
    - Following-up to ensure that remedial action has been effective
2. Inform building staff of these procedures.
3. Inform building occupants and/or tenants of these procedures and periodically remind them how to locate responsible staff and where to obtain complaint forms.

Reproduced from US EPA & NIOSH, Building Air Quality Action Plan, June 1998.

ENDNOTES

1. Carroll, Lewis. *Alice's Adventures in Wonderland.* A facsimile edition of the First Edition. New York; 1941.
2. WHO. *Guidelines for Air Quality.* World Health Organization, Geneva, Switzerland; 2000.
3. EPA. *Indoor Air Pollution: An Introduction for Health Professionals.* US Government Printing Office 1994-523-217/81322. Published jointly by American Lung Association, American Medical Association, US Consumer Product Safety Commission, US Environmental Protection Agency; Washington, DC, 1994.
4. ANSI/ASHRAE Standard Number 62-1989, *Ventilation for Acceptable Indoor Air Quality.* ANSI, The American National Standards Institute; ASHRAE, The American Society of Heating, Refrigeration, and Air Conditioning Engineers.
5. EPA/NIOSH (June 1998). *Building Air Quality Action Plan.* US Environmental Protection Agency/National Institute of Occupational Safety and Health. EPA Publication 402-K-98-001. DHHS (NIOSH) Publication No. 98-123.

chapter 13

Patient Safety Program

Lessons from other industries teach that we can do a much better job in capturing data about the things that almost went wrong—and use it as a predictive tool in avoiding serious injuries.

Martin J. Hatlie, JD, Executive Director, American Medical Association

An elderly woman with a vision problem and dependent on a walker is escorted into a medical practice by her husband, son, and daughter. The husband trips over a curb, falls forward into a trash can, and injures his leg. His wife becomes distraught, stumbles and falls, and strikes her head on the walker. The son and daughter call for help and the husband and wife are admitted to the triage area. Later, the wife is admitted to the hospital and the husband is on crutches. How could this happen at a facility for the sick and injured?

Homes are often childproofed, especially when young children are present. Childproofing means going around and identifying what can harm the child and then removing the object, putting it in an inaccessible place, attaching a lock to a door to prevent access, blunting a sharp corner, or removing a slippery throw rug that may cause unsteady feet to slip or trip. Essentially, a safety hazard evaluation and risk assessment is performed on the home.

A medical practice is subject to the same safety problems that other industries encounter. There are slip, trip, and fall hazards; electrical hazards; scalding hazards; exiting hazards; weather and utility hazards; fire and smoke hazards; construction hazards; chemical hazards; biological hazards; lighting hazards; equipment hazards; procedure hazards; and hazards that are specific to the staff who work in a medical practice.

Patients also present potential unique safety hazards, such as being ill; preoccupied with an illness, vision, or hearing impairment; or unsteady when walking.

OSHA requires employers to provide a workplace that is safe from known and potential safety and health hazards that may cause injury, illness, or even death. Although OSHA requirements apply to employees, if the medical practice is required to be safe and healthy for the staff, then it should also be a safe and healthy place for patients, visitors, and families.

This book uses a basic process of recognize, evaluate, and control to help employers thoroughly embrace and implement the concept of safety. This development process should also be used for the patient safety

TABLE 13.1

Checklist for Developing a Patient Safety Program

Administrative Support	Allocate time, funds, and people
Hazard Identification	Conduct walk-through inspections and identify hazards
Hazard Evaluation	Determine needs
Training	Provide training to employees
Controls	Select and implement control measures
Program Review	Review program to measure success
Recordkeeping	Maintain records of inspections, evaluations, corrective actions, success, and so forth

Reproduced from NIOSH, *Guidelines for Protecting the Safety and Health of Health Care Workers*, 1998.

program. The NIOSH document, titled *Guidelines for Protecting the Safety and Health of Health Care Workers*,[1] introduced the concept to help health care facilities address their diverse safety and health concerns. The components of the NIOSH document acknowledge OSHA's requirements and can help carry those requirements forward through the program development process. Table 13.1 summarizes the NIOSH concept that can be used for the patient safety program.

FACILITY AND ENVIRONMENT

A health care provider must make the health and well-being of patients a primary concern. That concern includes making their visit to the office not only a pleasant experience but a safe one as well.

In many cases, the medical practice is located in space that is leased, so there are permanent staff employed by the building owners who provide building maintenance. In other cases, the practice owns the building and must provide building maintenance.

One simple and effective way to identify potential safety hazards in and around the medical practice is to conduct a visual inspection of the property. When conducting this safety tour, try to imagine what extremely ill patients might have to go through to reach the doctor's office. If there is anything that could obstruct their passage, be unreliable (such as a loose handrail), or possibly impede their progress, it must be identified, fixed, or managed.

If young children are frequently present at a practice, survey the area to see what they will touch, feel, or taste. Dangling cords, unprotected wall outlets, and other dangers might catch a child's eye, leading to a serious safety hazard. There are two areas that require attention: the outside of the facility and the inside.

Outside the Facility

Take on the role of an arriving patient and take an "arrival tour" during the day as well as after dark. In many parts of this country, it is already dark by 4:00 or 5:00 pm during the winter months. The northern states are subject to extremes in temperature, which means surfaces contract or expand. Frost can play havoc with a parking lot in a relatively short period of time. A heavy snowfall can overburden tree limbs until they snap and fall, possibly causing injury.

Parking Area

Pay attention to the location of the entry and exit driveways, the flow of traffic through the parking lot, the lighting, and the physical condition of the parking area. Ask questions such as the following:

- Can I enter and exit without fear of striking or being struck by another vehicle?
- Can I easily negotiate the parking aisles?
- Are the parking spaces easy to enter and exit?
- Are the parking spaces clearly visible?
- Is the lot well-lighted?
- Are there potholes?
- Are handrailings secure or loose?
- Do water and ice collect in the parking lot or the walkways?
- Are lawns and landscaping free from debris?
- During cold winter months, are icicles and slabs of snow cleared when necessary so they can't fall onto walkways?
- Are there handicap-accessible parking spots?
- Are the handicap-accessible parking spots accessible?

Sidewalks, Ramps, and Stairs

When walking to the building, pay attention to the sidewalks, ramps, and stairs.

- Are the sidewalks, ramps, and stairs in good repair?
- Are they finished with a surface that helps prevent slipping?
- Are they wide enough for two people walking side by side?
- Are they free from overhanging shrubs and tree limbs?
- Are handrails secure on stairs and ramps?
- Are steps in good repair with smooth surface?
- Are safety treads in good condition?

Building Entrance

When entering the building, ask the following questions:

- Do the automatic and nonautomatic doors open easily?
- What is the condition of the threshold doorstep?
- Is the building handicap-accessible?
- Can a person on crutches, in a wheelchair, or using a walker or cane easily enter the building?
- Are the walkways (including ramps) in good condition?
- Are the entrances well-marked and lighted?
- If there are stairs, are the handrails in good repair and are the steps wide enough to allow comfortable walking?
- Are the overhead ceiling tiles clean and in place?
- Are the lighting fixtures secure and properly functioning?

- Do the exit lights function, point in the correct direction, and are they easy to see?
- Are there a number of sheltered benches available for those who are waiting for transportation?

Inside the Facility

When entering the front door of the medical practice, bear in mind the possible safety issues that may face the patient, and take corrective action.

Office Entrance

The entrance to the medical practice is well traveled. To be sure safety issues have been addressed with regard to the office entrance, ask the following questions:

- Are the vestibule and other patient areas well-lighted?
- Are the floors finished with a solid surface that allows patients in wheelchairs to easily move throughout the space?
- Are the floors free of trash and obstacles?
- Do elevator doors allow ample time for slower-moving patients or those in wheelchairs to enter and exit?
- Do elevators operate smoothly?
- Is there a separate service elevator?
- Are pictures or other decorative items placed high enough on the walls in order to prevent children and adults alike from being injured?
- Are hallways easy to negotiate either on foot or in a wheelchair?

Waiting Area

Patients spend a fair amount of time in the waiting area, so ask the following safety questions about this important room:

- Is the lighting adequate?
- If there are lamps, is the excess cordage out of the way to prevent tripping?
- Are electrical outlets that are not in use childproofed with outlet protectors?
- Is the furniture in good repair?
- If there is a children's play area, is it away from the main traffic pattern?
- Are all toys and child-size furniture in good condition and free of sharp edges, cracks, or other possibly harmful conditions?
- If the building has windows that open, are there barriers in place to prevent falls or objects being dropped out the window?
- If there are potted plants (real or imitation), are they in wide-based containers and located out of the patient's way?
- Are there wastebaskets conveniently located throughout the area?

Hallways and Aisleways

Many times various things are left in the hallway. From cleaning materials to empty or unopened cartons to almost anything else, it often ends up in a hallway. *Nothing* should be left in a hallway that doesn't absolutely belong there. Inspect hallways for possible hazards with the same thoroughness as other areas by asking the following questions.

- Are all lighting fixtures in good working condition and flickering or burned-out bulbs replaced?
- Are handrails, if any, clean (to avoid losing one's grip) and firmly installed, with no wobbling?
- Are floors clean and free from litter?
- If the offices are large enough to have connecting hallways, are convex mirrors installed at corners to avert accidents?
- If doors open out into a hallway, are convex mirrors installed to prevent an accident?
- Are electrical outlets that are not in use childproofed with outlet protectors?
- Are spillable containers that are transported down a hallway securely fitted with a lid?
- Are the exit lights operational, readily seen from at least two directions, and with directional arrows, as appropriate? Are the emergency batteries in the exit lights tested at least monthly?
- Are all exit doors free of obstruction and well-marked?
- Are hallway doors that lead to other areas (eg, storage rooms, mechanical rooms) labeled and locked, if appropriate?
- Is the emergency lighting tested monthly?
- Do filing drawers block aisleways?

Preexamination Room

In some practices, there is a room or alcove where the patients are seen before being shown to the examination room to await the doctor. Generally, the weight, temperature, and blood pressure of the patient are taken and recorded in this room. When patients are in good health, this simple preexamination procedure is straightforward and uncomplicated. However, when dealing with patients who are extremely ill, physically challenged, children, or senior citizens, other considerations must be taken into account. In these instances, the patients may have difficulty standing on a scale. It is very important to make sure the nurse can handle the patient's weight should the patient begin to fall. In all likelihood, these patients were brought to the medical practice by a friend or relative. Whenever possible, invite them to be present with the patient. In certain instances, these patients may feel disoriented or even frightened and possibly try to leave the area; having someone they know nearby is often reassuring. Being prepared for all contingencies greatly aids in maintaining patient safety.

Children bring a different set of possibilities by virtue of their unending curiosity. The best way to preserve the safety of children is to protect them from themselves. Medical instruments must be kept out of their reach. Medical supplies should be kept in a cupboard or closed drawer.

It may be prudent to invest in closures that are magnetized or in some other way offer a bit of resistance. The best rule of thumb is to use only what is needed at the moment, then immediately put it away. By doing so, the chances for injuries to patients are greatly diminished.

Examination Room

Examination rooms present special problems that should be addressed by first looking at who occupies them. For example, if there are pediatric and curious young adult patients, view the examination rooms the same way a home is childproofed when children are small. Because patients often have to spend time in the examination room waiting for the doctor, it is a good idea to have several magazines available. Distraction can go a long way to preventing opening of doors and drawers out of curiosity.

In an orthopedic surgeon's office, many patients are on crutches or in wheelchairs, and patients often wear splints and casts. Providing ample space for injured or immobilized limbs is imperative for safe maneuvering within the waiting area, hallways, or the examination room.

The exam table, especially because it usually has moving parts, must be thoroughly checked for safe functioning and smooth operation. Moreover, a careful inspection of the table's covering (usually a Naugahyde-type fabric) must be carefully gone over to be sure there are no rips, tears, holes, or any other sources of possible injury or a means of harboring germs.

The following is a checklist to help ensure a safe examination room:

- Do the cabinets and drawers have safety latches?
- Are drawers and cupboards locked when appropriate?
- Have outlets been childproofed with outlet protectors?
- As in other areas, is it easy for handicapped patients to navigate about the room?
- Are syringes, scissors, or other sharp objects stored in overhead cabinets?
- Are surfaces (ie, sides and bottoms as well as tops) smooth with no worn or broken areas that could lead to splinters, scratches, or other injuries?
- Are mirrors mounted on a wall, and not on the door?
- Is medical equipment that relies on hoses, tanks, or canisters routinely checked for operational safety (eg, no clogs, frays, or holes in hoses)?
- Is electrical equipment checked as required and cords inspected for frays or splits?
- Is there some type of signaling device (eg, hand bell, buzzer) for a patient needing immediate assistance?

Storage Areas

Medical practices have storage, which should have appropriate locking devices to guard against anyone mistakenly entering the closet and possibly being injured. Storage closets hold cleaning products, mops, vacuums, and so forth.

There should be a separate locked area with suitable shelving for fundamental supplies such as unopened boxes of disposable gloves, paper towels, toilet tissue, or any other items that are routinely used in a medical practice.

Pharmaceuticals, medications, and controlled drugs must be stocked and stored using tracking systems that comply with required inventory control standards.

Lavatories

Sometimes the lavatories are located within the medical practice's offices, and sometimes they are a common facility found in a hallway that serves an entire floor in the building. If a hazard or potential hazard is identified in a lavatory outside the medical office, the building's maintenance staff must be notified of hazard, safety, and sanitation concerns. Whether within the practice's space or down the hallway, there should be a signaling device for patients needing immediate help (eg, air horn, hand bell, buzzer).

Frequent inspections for cleanliness and safety are extremely important. Wet and slippery floors, excess toilet tissue on floors, and anything that diminishes the sanitation and safety inside a lavatory needs to be attended to as quickly as possible. Soap dispensers and paper towels must be provided and refilled or replaced as needed. Handwashing may lead to spilled water or soap on the floor, making it very slippery. Grab bars or safety bars may need to be installed. If they are installed, they must be routinely checked for security. Damage, including bends and cracks, must be identified and the grab bars replaced.

Exits

In large practices, often with a confusing, maze-like floor plan, the way back to the lobby or exit area must be clearly indicated. When patients enter the exam room, they are usually escorted there by a nurse. However, it is not customary for patients to be escorted back to the waiting room or exit area.

If patients are expected to go to a different area to present health insurance information, make payment, or set a new appointment, make sure the route is clearly marked. It is easy for patients to become disoriented. Color-key signs, pictorial signs, or directional maps help navigate down hallways and through rooms. Signage systems with multilingual capability should be considered in certain geographic areas of the country. Americans with Disabilities Act (ADA) compliance and NFPA Life Safety issues must be implemented. In addition to safety and ADA considerations, effective signage is an important public relations tool for the medical practice.

ADDITIONAL CONSIDERATIONS

One of the best ways to provide patient safety is to observe good housekeeping practices. This precaution increases the awareness of daily hazards that become obvious and often are ignored. Clutter is the culprit in many incidents involving patients on crutches, wheelchairs, or using walkers or canes. Poor lighting or uneven floors are also a hazard to patients who must rely on some form of support for mobility.

Senior Citizens

When referring to senior citizens, it is assumed that they are in their advanced years, when the body's organs do not work as well as they used to, and when falls can be infinitely more serious and injurious. While some people can live well into their 90s and never know a serious day's illness, they are in the minority. For the rest, life can be living with painful arthritis, debilitating allergies, or a host of other complications that the body can no longer ward off.

These patients may need greater assistance in the waiting room, louder and clearer communication, and speedier, friendly guidance from one room to another so that they do not become disoriented. Although these patients may come to their appointment with someone to assist them, this is not always the case. Employees should be alert to any possible hazards involved.

Falls

Falls are a serious public health problem among older adults. In the United States, one of every three people 65 years and older falls each year.[2,3] Older adults are hospitalized for fall-related injuries five times more often than they are for injuries from other causes.[4] Of those who fall, 20% to 30% suffer moderate to severe injuries that reduce mobility and independence, and increase the risk of premature death.[4] Chances of falling increase if the person has problems with gait or balance, poor distance vision, or gets light-headed when standing up. If the person is taking sedatives or tranquilizers, there is an increased risk of falling. Such grim statistics become more serious given the aging US population. Already the over-85 age segment is the fastest growing, and within a decade the post-World War II baby boomers will begin entering their golden years. By 2030, 20% of the population will be over 65, up from 13% in 2001.[5]

To reduce the risk of senior citizens falling while at the medical practice, the following includes special supportive features to help prevent falls among the elderly:

- Increase lighting in the office environment, especially stairwells and bathrooms.
- Keep power cords and decorative items off the floor.
- Nail down area rugs or get rid of them and replace flooring that warps or buckles.
- Install a toilet seat that is built up higher than normal.
- Make sure handrails on staircases are secured.
- Create color contrasts between walls and floors; lighter-colored floor surfaces are preferable.
- Install slip-resistant tile in bathrooms.
- Minimize changes in walking surfaces, and use slip-resistant coverings, such as rough tile and carpet with short, dense pile.
- Position coatracks so they are not too high or behind a door where a fragile, elderly person could be knocked over when someone else comes into the medical practice.
- Examination rooms should be large enough to maneuver a walker or wheelchair.

- Invest in a motorized exam table that will go low enough so the elderly person does not have to climb up on a standard gynecologic table or use a step stool. It may be possible to examine the elderly patient while they sit in a chair or wheelchair.
- Invest in a scale with handles so senior citizens can hold on for balance. The standard type of weight scale presents difficulties for elderly people, who may have unsteady balance.

Senior citizens who have fallen or have seen their friends fall may develop a fear of falling, which can significantly affect their quality of life and threaten their ability to live independently.

Children

When children frequent a medical practice, they bring with them a vast array of safety issues and concerns that are unique to their age group.

Toy Safety in Waiting Areas

Toys can be a source of fun, learning, and entertainment for children. In medical practices, especially pediatric offices, play areas for children are important. However, toys with poor design, toys that are too old for a child's age or are used incorrectly, or toys in bad repair can lead to serious, even fatal, injuries. How can medical practices be sure that toys are safe to bring into the children's waiting areas? It is not always easy to tell. Some safety standards, like the prohibition of the use of lead paint or insulation standards for electric toys, are not always apparent to the eye. Most toys are packaged in ways that make it difficult for the purchaser to check. And most consumers do not know what to look out for when toy shopping. Despite passage of the 1994 Child Safety Protection Act, hazardous toys can still be found on toy store shelves across the country and can make their way into medical practices.[6] Such hazardous toys are required to carry the act's warning on their packaging, as illustrated in Figure 13.1.

The US Public Interest Research Groups (PIRG) is the national lobbying office for state PIRGs and conducts toy safety surveys. It uses the results from its surveys to educate about toy hazards and to advocate passage of stronger laws and regulations to protect children from toy hazards. The following tips for toy safety include information from PIRG's 1999 Report on Dangerous Toys, *Trouble in Toyland*[7]:

- When purchasing toys for children, check for small parts that may choke or cords that may strangle.
- Toys with small parts, small balls, and marbles are banned for sale if intended for children under 3.
- Obtain a *no-choke testing tube* to see if a toy or toy part is potentially dangerous because of its size. Commercial tubes are available in better toy stores. An empty toilet paper tube is also effective.

FIGURE 13.1

Hazard Warning Required on Toys Intended for Children 3 to 6 Years Old.

Reproduced from Public Interest Research Groups, *PIRG Tips for Toy Safety.* Copyright 1999.

Warning: CHOKE HAZARD: Small Parts. Not for Children Under 3.

- Be conscious of objects that have potentially dangerous small parts: removable eyes and noses on stuffed toys and dolls; small, removable squeakers on squeeze toys; and little figures and pieces that fit into larger toys.
- Latex balloons are the leading toy killer. If a balloon burst while a child is blowing it up, it could be inhaled. Supervise children with latex balloons, inflated or not. Consider foil balloons because they do not present the same kinds of hazards to young children.
- Children as old as 5 have choked to death on small balls and marbles as large as 1.75 inches. Be careful of ball-like beads and other round objects.
- Strings, cords, and necklaces can strangle infants.
- Projectiles (ie, flying dolls and action figures that catapult off handheld launchers) can lacerate skin and blind or deafen a child who is struck.

Many guidelines, recommendations, and regulations are available to help people purchase safe toys. The March 1997 policy statement by the American Academy of Pediatrics[8] can serve as a guideline. In addition, under the Federal Hazardous Substances Act (FHSA) and the Consumer Product Safety Act (CPSA), the Consumer Product Safety Commission (CPSC) regulates certain toys. The CPSC works to reduce the risk of injuries and deaths from consumer products as outlined in Table 13.2. CPSC regulations address numerous other toy-related hazards including the following information regarding toys that could be found in the waiting rooms of medical practices:

- Toys with sharp points or edges can lacerate, cut, or puncture skin.
- Rattles must be large enough so that they cannot become lodged in an infant's throat. Rattles must also be constructed so that they will not separate into small pieces. Be wary of rattle-shaped toys, such as xylophone mallets, that may not meet the rattle test because they are intended for older children.
- To prevent lead poisoning, the amount of lead in paint used on toys and other children's articles is limited to less than 0.06%. In the last several years there has been a sharp increase in the number of toys recalled by the CPSC for excess lead in paint.

Younger kids often put toys in their mouths, and having toys prone to go in kids' mouths is a very easy way for germs to spread. When tidying up

TABLE 13.2

Consumer Product Safety Commission Responsibilities

- Developing voluntary standards with industry
- Issuing and enforcing mandatory standards; banning consumer products if no feasible standard would adequately protect the public
- Obtaining the recall of products or arranging for their repair
- Conducting research on potential product hazards
- Informing and educating consumers through the media, state and local governments, private organizations, and by responding to consumer inquiries

Reproduced from Consumer Product Safety Commission.

the play area, be alert for toys that were left behind by children or parents and discard them, as they could later pose a hazard to a younger child and be unsanitary.

Childproofing the Waiting Areas

As with the other areas of the medical practice, the waiting area(s) for children should be clean, safe, and orderly. This can be a difficult task in busy medical practices. The following are suggestions to help make a potentially formidable task easier:

- Children are attracted to electrical outlet plates with open prong holes. The outlet plates should be covered using traditional outlet protectors or other commercial devices on the market.
- Cover cabinet doorknobs with plastic knob covers that can be cleaned and sanitized.
- Bolt bookcases, cabinets, or any tipable furniture to the wall or floor with L-braces or hook-and-eye hardware. Commercial furniture braces are also available and attach easily to the back of furniture.
- Cords should be hidden, not loosely lying where little ones may trip, tug, or chew. Cords can be placed inside Velcro strips along baseboard or walls.
- Children can be protected from furniture with sharp corners and edges with padded shields. Nontoxic shields that slip easily over the sharp corners and edges of most coffee tables can be purchased.
- Many indoor (and outdoor) plants are poisonous, including popular holiday plants such as the poinsettia. If poisonous plants are included in the office décor, they must be kept out of the reach of children.

Make sure that parents know that it is their responsibility to supervise their children in office play areas.

Persons with Mental Disabilities

Because of the wide range of possibilities—from mild mental disability to severe—common sense is the best approach. It is unlikely that a patient with a severe disability will be at a medical practice alone. A relative, friend, or perhaps a private-duty nurse will usually be with the patient. In some instances, helping the patient may require physical assistance from the individual; in others, to serve as a "translator" and be a comforting presence.

Other types of mental impairment may be as a result of drug or alcohol misuse or abuse. In some situations, the patient is no longer able to effectively communicate, either listening or speaking. Not only is more patience required, but also assistance in following instructions may be necessary. If vertigo (loss of balance) is part of the disability, it is all the more important to be sure that the staff is able to assist the patient.

Persons with Physical Disabilities

The ADA has made many people aware of the special problems that can be encountered by persons with physical disabilities. The act has also caused many businesses to make modifications to their facilities in order to comply with the ADA's regulations. While some have found the requirements of the ADA to be burdensome, the ADA requirements provide an additional measure of safety for patients.

EMERGENCY MANAGEMENT

Fires and other emergencies claim the lives of hundreds of Americans each year, injure thousands more, and cause billions of dollars worth of damage. Medical practices present a unique challenge in preparing for emergencies. Staff must protect the patients, while not placing themselves in undue danger. Evacuation must be secondary to a strong program of prevention.

MEDICAL EQUIPMENT MANAGEMENT

Effective medical equipment management programs help promote a safe patient environment. A functional program can help enhance the quality of patient care and prevent patient injury through early recognition of potential equipment problems.

Programs should include written policies and procedures, compliance with regulatory agencies, staff training, equipment inventory and documentation, and equipment safety and performance testing. Programs should also address equipment hazard notices and recalls, monitoring and reporting device-related injuries or deaths under the Safe Medical Devices Act of 1990, and reporting and investigating equipment management problems, failures, and user errors.

NATIONAL PATIENT SAFETY FOUNDATION

As the issue of patient safety gains momentum around the country, foundations and organizations are forming or gaining an increased presence in the marketplace. The JCAHO is continually working to improve patient safety.

Within the past few years, the Annenberg Center for Health Sciences, JCAHO, and the American Association for the Advancement of Science came together to explore the dimensions of the patient injury problem and its potential solutions. This effort launched the National Patient Safety Foundation (NPSF), an independent, nonprofit research and education organization. Table 13.3 provides the mission of the NPSF, which is to improve measurable patient safety in medical practices.

TABLE 13.3

The Mission of the National Patient Safety Foundation

- Identify and create a core body of knowledge
- Identify pathways to apply the knowledge
- Develop and enhance the culture of receptivity to patient safety
- Raise public awareness and foster communications about patient safety
- Improve the status of the Foundation and its ability to meet its goals

Reprinted with permission from National Patient Safety Foundation, published by the American Medical Association. Copyright 1995-2000.

Although improved patient care is the primary focus of the NPSF, the foundation researches and explores the similarities between health care and other sectors where there also are elevated risks of human injury. NPSF embraces the broad scope of the phrase, *A Culture of Safety.* Like worker safety, patient safety requires constructive thinking, compassion, program commitments, and accountability. Like other industries, patient safety includes people, equipment, and the environment, not just patient-related interventions. The foundation is working toward breaking new ground in understanding how "people create safety," whether these people include patients coming to a medical practice for treatment or employees at work taking care of the patients.

Like worker safety, patient safety emerges from the interaction of the components of the system; it does not reside in a person, device, or department. Improving safety depends on learning how safety emerges from the interactions of the components.

ENDNOTES

1. National Institute for Occupational Saftey and Health (1988). *Guidelines for Protecting the Safety and Health of Health Care Workers.* US Department of Health and Human Services, National Institute for Occupational Safety and Health, Cincinnati, OH. DHHS Publication No. 88-119.

2. Tinetti ME, Speechley M, and Ginter SF. Risk factors for falls among elderly persons living in the community. *New England Journal of Medicine,* 1988:319(26),1701-7.

3. Sattin RW. Falls among older persons: A public health perspective. *Annual Review of Public Health,* 1992:489-508.

4. Alexander BH, Rivara FP, and Wolf, ME. The cost and frequency of hospitalization for fall-related injuries in older adults. *American Journal of Public Health,* 1992:82(7);1020-3.

5. Rollins, G. *Preventing the Fall: Designs on Building Safe Homes for the Elderly.* Safety and Health, National Safety Council; Itasca, IL. September 2000.

6. Public Interest Research Groups (PIRG). *PIRG Report Finds Hazardous Toys.* PIRG Toy Safety Press Release.

7. Public Interest Research Groups (PIRG). *Trouble in Toyland: 1999 Report on Dangerous Toys*; Washington, DC.

8. American Academy of Pediatrics. (RE9714). *Policy Statement. Selecting Appropriate Toys for Young Children: The Pediatrician's Role (RE9714).* Committee on Early Childhood, Adoption, and Dependent Care; Elk Grove Village, IL. Available at: www.aap.org/policy.

chapter 14

Emergency Management and Response

Effects of a disaster can linger long after its occurrence, rekindled by new experiences that remind the person of the past traumatic event. A rainstorm can become the reminder of a flood; the flash of lightening and the crash of thunder, the reminder of an explosion; and a small earth tremor, the reminder of a major earthquake.

Robert J Ursano, MD, et al, Psychiatric Dimensions of Disaster[1]

Every year disasters take their toll on workplaces across the nation—in lives and dollars. Disasters take many forms. Winds, floods, fire, ice, and snow can cause tragedy. Arson and other acts of terrorism are unique disasters that can strike at any time. Some catastrophes give warning, like a storm preceding a flood, while others, like earthquakes or fires, give no warning at all. Once a disaster happens, the time to prepare is gone and all people can do is cope. When disaster strikes, the best response is knowing what to do.

Emergency management programs include steps to handle disasters in the workplace. These programs include fires, bomb threats, natural disasters, technological emergencies, and workplace violence. Medical practices need to be prepared for emergencies, not only for the safety of the employees, but for the patients in their office as well. Part of the training should cover dealing with frightened children, persons with physical or mental disabilities, senior citizens, or other special patient groups.

EMERGENCY MANAGEMENT CONCEPTS

Emergency management includes any activities that prevent an emergency, reduce the chance of an emergency happening, or lessen the damaging effects of unavoidable emergencies. The entire management process is quite dynamic. Planning, though critical, is not the only component. Training, conducting drills, testing equipment, and coordinating activities with the community are other important functions.

There are many negative effects of an emergency (eg, deaths, injuries, fines, criminal prosecution), but medical practices can experience numerous positive aspects of preparedness, such as:

- Helping medical practices fulfill their moral responsibilities to protect employees, the community, and the environment.

- Facilitating compliance with regulatory requirements of federal, state, and local agencies.
- Enhancing a practice's ability to recover from financial losses, regulatory fines, loss of market share, damages to equipment or products, or business interruption.
- Reducing exposure to civil or criminal liability in the event of an incident.
- Enhancing a practice's image and credibility with employees, patients, suppliers, and the community.
- Possible reduction of insurance premiums.

The Planning Process

The four steps in the planning process are: establishing a planning team, analyzing capabilities and hazards, developing the plan, and implementing the plan.

Establishing a Planning Team
There must be an individual or group in charge of developing the emergency management plan. If the medical practice is big enough, a group is best because it encourages participation and enhances the visibility and stature of the planning process.

Analyzing Capabilities and Hazards
In this step, information about current capabilities and possible hazards and emergencies is gathered. It also includes conducting a vulnerability analysis of the facility—the probability and potential effect of each emergency.

- Historically, what types of emergencies have occurred in the community, at this facility, and at other facilities in the area (eg, fires, severe weather, terrorism, bomb threats)?
- Geographically, what can happen as a result of the facility's location? Keep in mind nearness to flood plains and dams and proximity to earthquake- or tornado-prone areas.

Developing the Plan
The plan should include basic components, such as response procedures, support documents, training requirements, and insights and information from outside organizations. This is the step where all of the components are put in place.

Implementing the Plan
In this step, recommendations made during the vulnerability analysis are acted upon, integrating the plan into company operations, training employees, and evaluating the plan. Look for opportunities to build awareness, educate and train staff, test procedures, and make emergency management part of what personnel do on a day-to-day basis. Test the plan by asking if personnel know what to do in an emergency, what kinds of safety posters or other visible reminders would be helpful, or if there are opportunities for distributing information through corporate newsletters or employee mailings.

Emergency Management Core Considerations

The following are core operational considerations of emergency management:

- Direction and control
- Communication
- Life safety
- Property protection
- Community outreach
- Recovery and restoration
- Administration and logistics

Direction and Control

Someone must be in charge in an emergency. Larger medical practices may have a team, while smaller practices may rely on a job position or combine responsibilities. In this chapter, OSHA's term, *Emergency Response Team,* is used to mean one or more individuals who are in charge in an emergency.

Communication

Communication is essential to any business operation. A communication failure can be a disaster in itself, cutting off vital business activities. Communication is needed to report emergencies, warn personnel and patients of the danger, keep families and off-duty employees informed about what is happening at the facility, and coordinate response actions.

In this era of high-tech electronics, it is easy to rely on interoffice systems, telephones, or other types of communication. There must be a backup system, such as voice communication, for reporting the emergency and communicating with the staff and patients.

Life Safety

Protecting the health and safety of everyone in the facility is the first priority during an emergency. One common means of protection is evacuation. In the case of fire, an immediate evacuation to a predetermined area away from the facility may be necessary. In a hurricane, evacuation could involve the entire community and take place over a period of days. Only properly trained and equipped professionals should conduct search and rescue. Death or serious injury can occur when untrained employees reenter a damaged or contaminated facility.

In the event of an emergency, staff and patients need to be instructed how to safely evacuate. Largely, the type of premises determines evacuation routes where the practice is located. The fire department or other emergency personnel may be able to assist in establishing safe procedures.

Some types of illnesses (eg, asthma) may impede a patient's safe escape. Staff members must know how to assist those patients. Patients in wheelchairs or on crutches will also need assistance. During emergency preparedness training, the evacuation procedure should be rehearsed. The staff must understand evacuation procedures and how best to shepherd patients to safety.

Property Protection

Protecting facilities, equipment, and vital records is essential to restoring operations once an emergency has occurred. Vital records may include financial and insurance information, patient databases, and personnel files. Establish procedures for fighting fires, shutting down equipment, or containing material spills. Train employees to recognize when to abandon the effort. Obtain materials to carry out protection procedures and keep them on hand for use only in emergencies. Consider ways to reduce the effects of emergencies, such as moving or constructing the medical practice away from flood plains and fault zones.

Community Outreach

The medical practice's relationship with the community will influence the ability to protect personnel and property and return to normal operations. Maintain a dialogue with community leaders, first responders, government agencies, community organizations, and utilities. In an emergency, the media are the most important link to the public. Try to develop and maintain positive relations with media outlets in the area. Press releases about facility-generated emergencies should describe who is involved in the incident and what happened, including when, where, why, and how.

Recovery and Restoration

Business recovery and restoration, or business resumption, goes right to a facility's bottom line: keeping people employed and the business running. Consider contractual arrangements with vendors in advance for post-emergency services, such as records preservation, and meet with the insurance carriers to discuss policies and coverage. Employees are a valuable asset and may need support after an emergency, such as flexible work hours or crisis counseling. Immediately after an emergency, take steps to resume operations. Account for all damage-related costs, protect undamaged property, conduct an investigation, and restore equipment and property.

Administration and Logistics

Maintain complete and accurate records to ensure a more efficient emergency response and recovery. Certain records may also be required by regulations or insurance carriers and may prove invaluable in the case of legal action after an incident. Document drills and exercises and their critiques. Maintain training records. Issue press releases. Maintain telephone logs.

HAZARD-SPECIFIC INFORMATION

There are many hazards in the workplace. This section will focus on fires, natural disasters, technological emergencies, and bomb threats.

Fires

Fire is the most common of all the hazards. Every year, fires cause thousands of deaths and injuries and billions of dollars in property damage.

Natural Disasters and Fires

A wide range of disasters occurs within the United States every year and has a devastating effect on the workplace. There are many types of fire-related hazards involved with natural disasters. Employees can greatly reduce the chances of becoming a fire casualty by identifying potential hazards and following safety tips as outlined below[2]:

1. Fire-Related Hazards
   - Gas, chemical, and electrical hazards may be present.
   - Appliances exposed to water can short and become a fire hazard.
   - Leaking aboveground gas lines, damaged or leaking gas or propane containers, and leaking vehicle gas tanks may explode or ignite, leading to disastrous fires.
   - Generators used during power outages can be very hazardous unless properly used and maintained.
   - Pools of water and even appliances can be electrically charged, resulting in a dangerous electrical fire.
   - Lightning associated with thunderstorms generates a variety of fire hazards, including electrocuting on contact, downed power lines, splitting trees, and causing fires.
   - Debris can easily ignite, especially if electrical wires are severed.
   - Alternative devices that are correctly or incorrectly used can create fire hazards.
   - Exposed outlets and wiring could present a fire and life safety hazard.

2. Safety Tips
   - Keep combustible liquids away from heat sources.
   - If the workplace has sustained flood or water damage and an employee can safely get to the main breaker or fuse box, turn off the power.
   - Look for and replace frayed or cracked extension and appliance cords, loose prongs, and plugs.
   - Never strike a match or a cigarette lighter. Any size flame can spark an explosion.
   - Have licensed electricians and other professionals check the workplace for damage.
   - Report downed or damaged power lines to the utility company or emergency services.
   - A backup system for a medical practice can facilitate a safe evacuation, provide adequate electricity to complete a procedure, and a host of other benefits. Generators can be dangerous when not properly operated, and the manufacturer's instructions and guidelines must be followed.
   - Flashlights with strong beams should be readily available in each room. The batteries and condition of the flashlights should be checked at least every 6 months. When flashlights are not in use, store them with the batteries in reverse polarity to make them last longer.

- Emergency lights can be installed as long as the batteries and light bulbs are checked for proper operation at least once a month.
- Never thaw frozen pipes with a blowtorch or other open flame. Use hot water or a *UL*-listed device, such as a hand-held dryer.
- Thoroughly clean up chemical spills and place containers in a well-ventilated area.
- Assume all wires on the ground are electrically charged.
- Check smoke alarms (and carbon monoxide detectors) every 6 months and replace batteries.

Arson

Arson is a serious national problem that annually robs, injures, and kills thousands. Arson causes 700 deaths and $2 billion in property damage each year. The Federal Emergency Management Agency (FEMA) and other organizations are initiating community action programs for arson prevention. There are many tools available for medical practices to help reverse this growing trend. The following are examples of a few effective tools:

- Reporting suspicious behavior.
- Keeping shrubbery trimmed so passersby and patrols can observe the entire building.
- Locking doors, as appropriate.
- Installing burglar and fire alarm systems.
- Keeping a list of all office key holders.
- Periodically changing the locks, if necessary.
- Using good housekeeping practices.
- Properly storing flammables and combustibles.
- Using licensed plumbers, electricians, and heating contractors to do needed repairs.
- Post *employee only* signs and *no solicitation* signs where appropriate.

The majority of arsonists are seeking attention or revenge. Be aware of individuals who may have become disgruntled and may wish to cause damage to the property. Alert local law enforcement and staff of acts of vandalism, as these often precede arson.

Typically, arsonists set a fire anywhere they believe themselves to be alone and unobserved. While it may seem unlikely that this would happen in a medical practice , it must not be discounted. Approach individuals who seem to be merely *hanging around*. Arsonists do not want to be recognized and will usually leave if confronted. Be suspicious of persons carrying containers in or around the building, as arsonists may be carrying accelerants in any type of container.

Arson awareness activities include educating employees to recognize behavior and developing a plan of action to put it into place, if needed.

Preventing Fires

Fire prevention involves identifying potential hazards, offering fire safety awareness, using common sense, utilizing good housekeeping practices, adhering to regulations for safe and proper storage of hazardous materials, enforcing the strict adherence to a smoking policy, and instilling the virtues of the correct use of tools and equipment.

Employees should take responsibility for a clean and tidy workspace. Trash containers should be emptied daily. Even the most seemingly insignificant things can lead to or promote a fire. For example, leaving paper trash lying around in work and storage areas, storing flammable liquid containers uncapped, or using electrical equipment with frayed cords can all contribute to unsafe work conditions.

Flammable liquids are another potential fire source. They must be stored in accordance with local fire regulations. MSDSs also provide use and storage information to avoid a fire. Training in a medical practice should include learning what flammables are used within the practice and proper storage and use precautions.

Part of daily observations includes being mindful of the smell of tobacco smoke in the air—from a restroom, storage closet, stairwells, or other types of sequestered areas. Monitor the patient waiting area, alcoves, restrooms, or similar areas where smokers may be out of sight, but not "out of smell."

Careless smoking can cause fires. Smoking policies exist for the comfort and well-being of people and to eliminate potential fire hazards.

Fires involving the use or misuse of electrical equipment are the number one cause of fire in the workplace. Proper application of electrical safety requirements is essential to safeguard workers from the dangers of fire, and personnel must be trained to take precautions. Checking electrical equipment for old or worn wiring and reporting hazardous conditions for repair can prevent fires. Wall sockets should not be overloaded, and one outlet should have no more than two plugs. Appliances or equipment that are plugged in and smell strange are often the first sign of a fire and should be investigated. Unlike an ordinary office, a medical practice relies on many different types of special medical equipment that requires the use of electricity.

Fire Extinguishers

When employers provide portable fire extinguishers for use in the workplace, certain OSHA requirements apply. OSHA's Fire Protection Standard (Standard Number 29 CFR 1910.157) outlines the requirements, such as inspection, maintenance, testing, and employee training.

Meet with the local fire department to talk about the community's fire response capabilities as they relate to the operation of the medical practice. The local fire department or a fire protection equipment company can help in selecting the most appropriate extinguisher. Not all fire extinguishers contain the same ingredients. Each fire extinguisher displays a rating on the faceplate showing the class of fires (A through D) it is designed to put out. Some extinguishers are marked with multiple ratings, such as AB, BC or ABC. Table 14.1 provides the classes of fire and fire extinguisher ratings. Extinguishers should be selected based on the classes of potential workplace fires and on the size and degree of potential hazards. Fire extinguishers must be placed in appropriate locations and clearly marked with a fire extinguisher sign. Table 14.2 provides a fire extinguisher checklist.

TABLE 14.1

Classes of Fire and Fire Extinguisher Ratings

Class A	Ordinary combustibles, such as wood, paper, cloth, rubber, or certain types of plastic.
Class B	Flammable or combustible gases and liquids, such as gasoline, kerosene, paint, paint thinners, or propane.
Class C	Energized electrical equipment, such as appliances, switches, or power tools.
Class D	Certain combustible metals, such as magnesium, titanium, potassium, or sodium.

Note: The facility may have halon extinguishers in the office. Although halon extinguishers are effective in extinguishing certain types of fires, halon has been shown to be harmful to the environment and is no longer being produced. Halotron 1 fire extinguishers are now available as a replacement for halon, and they are environmentally safe.

Reproduced with permission from *Central Training Technologies.* Copyright 1999. For more information, visit Coastal's website, www.coastal.com.

TABLE 14.2

Fire Extinguisher Checklist

- Proper type and size of extinguisher is selected to match the anticipated fire hazard.
- Extinguisher is properly mounted and stored in an obvious location near the fire hazard area, but not so close as to be potentially involved.
- Extinguisher's operation and use instructions are fully understood.
- Extinguisher is kept in a fully charged and serviced condition.
- Attend a training session or view a fire fighting video to better understand how to fight various types of fire configurations with a portable fire extinguisher.

Reproduced with permission from National Association of Fire Equipment Distributors, *Hand Portable Fire Extinguishers.*

Portable fire extinguishers are mechanical devices that require periodic service and maintenance to control optimum operating conditions. While disposable-type models cannot be serviced, they must be checked for periodic replacement requirements. Any time an extinguisher is discharged, it must be recharged or replaced.

Fire Response Training and Drills
Regular fire drills should be conducted to identify problems before an actual fire occurs. With an average of 6,000 office-building fires reported each year, it is essential employees know how to quickly and effectively respond to a fire emergency.

If the medical office is located on the ground floor or is located in a one-story building, and if the fire begins in the waiting room, setting carpeting, draperies, and all else into an inferno blocking access to the primary exit (ie, the front door), where is the alternate exit and how do staff and patients access this alternate exit? The chances for panic are strong, especially if the alternate evacuation routes and exits are not well marked and labeled.

What if a fire breaks out in a high-rise office building instead of a one-story structure? Because modern elevators are designed to lock down in such an instance, what measures will be in place to assist patients in wheelchairs or those too sick to walk down multiple flights of stairs? What if the stairwells are already crammed with screaming, panicked people blocking a safe exit?

Fire procedures in a one-story office may differ from procedures in a multistory building. The Emergency Response Team should establish if the building's management personnel have specific procedures in the event of a fire. In addition, local resources, such as the fire department, are often available to come on-site and conduct fire hazard surveys and provide employee training.

When conducting drills, every effort should be made to portray them as realistic as possible. Ideas include using kitty litter bags to carry like infants or putting full sandbags in wheelchairs to simulate adult patients.

Part of the practice's fire drills should include how to safely evacuate the ill or feeble patients who are unable to assist themselves. Persons who are very ill, feeble, or disabled should be placed in a wheelchair for evacuation. Depending upon the size of the medical practice, one or more wheelchairs should be available for evacuation and medical needs. If the medical practice is on an upper floor, then the practice should consider purchasing an evacuation chair and practice using it, going up and down the stairs. Another method used to evacuate from other than the ground floor is to use a sheet or blanket. During a fire drill, have an employee sit on a sheet or blanket and have another staff member practice evacuation by slowly pulling them down the stairs.

Practicing evacuation by crawling on hands and knees—or even elbowing one's way in a prone position—would be a realistic exercise. Drilling until it is an automatic habit to touch doorknobs (checking for heat on the other side) is also realistic. It is not enough to just *say* this is what must be done; it must be physically rehearsed, over and over, until the response training is done without thinking.

Using a Fire Extinguisher

Employees must know how to use a fire extinguisher, and training should include practice actually using this tool. One method to use during training includes pull, aim, squeeze, and sweep (PASS):

Pull the pin.
Aim the extinguisher at the base of the flames.
Squeeze the handles together, while holding the extinguisher upright.
Sweep the extinguisher from side to side, covering the area of the fire with the extinguisher agent.

Only use a fire extinguisher when it is safe to do so. If the fire is too big, or if it is spreading or threatening to block the path of escape, immediately leave the area. If necessary, use the extinguisher to clear an escape path.

What to Do in Case of Fire

In addition to fire prevention training, the Emergency Response Team must train employees on the proper response to a fire. Assistance may include aiding in the evacuation of patients or similar support.

Being prepared has saved many lives. Practicing what to do can help employees think and act quickly and safely, saving the lives of employees and patients. A code, such as "code red," may be used to announce the fire so that the staff can be alerted to its presence without frightening patients and risking a panicked rush to escape.

Acronyms, such as RACE, are often used to help teach employees critical responses and steps to take in the event of a fire:

> *Rescue anyone in immediate danger.* This is an important time for teamwork. At the same moment, someone else could be told to activate the alarm.
>
> *Activate the alarm or call 911 immediately.* Staff, patients, or visitors can continue to be evacuated while awaiting the fire department.
>
> *Confine the fire.* Be sure all windows and doors are tightly closed to prevent the spread of smoke and flames.
>
> *Evacuate or extinguish the fire.* Attempts to extinguish the fire should only be made if the fire is small and easily doused, such as a fire in a wastebasket. While smothering it—with a blanket or raincoat, perhaps—is one way to extinguish a fire, it offers the possibility of the fabric catching fire as well. The most preferable way to put out a fire is with a fire extinguisher.

Whether the procedural rules are for small medical practices or large ones, all rules must be consistently observed and practiced. In an emergency, it is easy to become disoriented or even briefly immobilized. Knowing the procedure by rote can make an enormous difference to the safety of employees and patients.

Evacuation

Few things are more frightening to people than fire, and people who are frightened often act irrationally. For this reason, it is imperative that the staff responds to a fire in quiet, collected manner.

When fire or power outages occur in a building, the degree of visibility in corridors, stairs, and passageways may mean the difference between orderly evacuation and chaos—possibly between life and death. If the exit or the way to reach it is not readily apparent to the staff or patients, approved signs that are readily visible from every direction must mark access to exits. The NFPA 101 Life Safety Code[3] and other NFPA standards and handbooks address exit requirements as well as other provisions relating to fire and life safety and should be consulted for code requirements. The OSHA Standard (Standard Number 1910, *Subpart E - Means of Egress*) also addresses exit requirements, such as the need for exiting routes, exits, exit lights, and emergency action plans.

If a fire is detected early enough, there should be ample time to calmly, but quickly, evacuate the building. Even with ample time, there are those who overreact—occasionally, even violently (eg, hitting, shoving). Staff must be taught how to handle such situations.

There may be patients who are too ill or injured to leave the premises without help from the staff. Because of their vulnerable condition, they may experience greater fear. It is important to soothe and reassure these patients and guide them to safety.

If a fire is not quickly confined, smoke inhalation is a major threat to personnel and patients. The last one out of a room should close the door, but it should not be locked. Locking the door can hinder the fire department's search and rescue efforts. If possible, employees and patients should cover their noses and mouths with a damp cloth to help them breathe.

Because smoke is warmer than air, it rises. The NFPA uses procedures, such as *stop, drop, roll, and cool*,[4] to teach school students how to survive fire by crawling low in smoke and when clothes catch fire, to STOP, DROP, AND ROLL to put out the flames and COOL the burn with cool water for 10-15 minutes and then get help if needed. This logic also applies for employees throughout workplaces, and it may also be necessary to instruct patients and visitors of various safety actions.

Elevators must never be used because a loss of power could trap people inside. Stairways must be kept free of materials so evacuation is not hindered or blocked. A staff member should be posted outside of the facility to ensure that once out, no one goes back inside and everyone is well away from the exit so people can safely leave. Once safely outside, employees should immediately report to a predetermined area for a head count.

The premises must also be checked carefully for children. Often, when frightened, children will hide under desks or tables, go into closets, or stand behind draperies. Parents must be instructed to be sure their children are with them, preferably holding their hands. Smaller children may have to be carried to safety.

If Trapped in a Burning Building

If trapped in a building that is on fire, do not panic. The ability to think clearly could save lives. If a telephone is available, call 911 and state the exact location. Never open a closed door without feeling the door first with the back of a hand. If the door is hot, try another exit. If no other exit exists, seal the vents and cracks around the door with anything available. If people are having difficulty breathing, instruct them to remain close to the floor and ventilate the room by opening or even breaking a window. If clothes catch on fire, remember to stop, drop, and roll—do not run. Running only feeds the fire with more oxygen, making it burn faster. If another person catches on fire, smother the flames by grabbing a jacket, blanket, or rug and wrapping it around them.

Surviving After the Fire

Recovering from a fire can be a physically and mentally draining process. When fire strikes, lives are suddenly turned around. Often, the hardest part is knowing where to begin and who to contact. FEMA provides the document, *After the Fire! Returning to Normal*,[5] and stresses that employers be aware of and educate employees in the following information and cautions:

- *The first 24 hours after a fire are critical*. The local disaster relief service and insurance company should be contacted.
- *Do not enter the damaged site without clearance from authorities*. Fire can rekindle from hidden, smoldering remains.

- *Do not attempt to turn on utilities.* Normally, the fire department will see that utilities are either safe to use or are disconnected before they leave the site.
- *Be watchful for structural damage caused by the fire.* Roofs and floors may be damaged and subject to collapse.
- *Food, beverages, and medicine exposed to heat, smoke, soot, and water should not be consumed.*
- *The owner of the medical practice should contact the police department to let them know the site will be unoccupied.*
- *Do not throw away damaged goods until an inventory is made.* All damages are taken into consideration in developing the insurance claim.

There are companies that specialize in the restoration of fire-damaged structures. The owner of the medical practice should become familiar with the types of services, be clear about who will pay, obtain estimates of cost, and check references. Restoration companies provide a range of services that may include some or all of the following:

- Securing the site against further damage
- Estimating structural damage
- Repairing structural damage
- Estimating the cost to repair or renew items of personal property
- Packing, transporting, and storing of office items
- Securing appropriate cleaning or repair subcontractors
- Storing repaired items until needed

Some items may be salvageable, including equipment, carpet, books, appliances, walls, floors, and furniture. Damage must be inventoried and assessed, and recommended cleaning or servicing procedures must be followed. Table 14.3 provides a quick reference and guide to follow immediately after a fire strikes.

TABLE 14.3

Quick Reference Guide to Follow After Fire Strikes

1. Contact the insurance company for detailed instructions on protecting the property, conducting inventory, and contacting fire-damage restoration companies.
2. Check with the fire department to make sure the facility is safe to enter. Be watchful of any structural damage caused by the fire.
3. The fire department should see that utilities are either safe to use or are disconnected before they leave the site. Do not attempt to reconnect utilities.
4. Conduct an inventory of damaged property and items. Do not throw away any damaged goods until after an inventory is made.
5. Locate valuable documents and records.
6. If the site will be unoccupied, inform the local police department.
7. Begin saving receipts for any money spent related to fire loss. The receipts may be needed later by the insurance company and for verifying losses claimed on income tax.
8. Check with the accountant or the Internal Revenue Service about special benefits for people recovering from fire loss.

Reproduced from Federal Emergency Management Agency, United States Fire Administration, *After the Fire—Returning to Normal.*

Natural Disasters

Natural disasters include earthquakes, floods, summer and winter storms, tornadoes, and hurricanes. Mitigation activities can lessen the damaging effects of natural disasters and can take many forms. For example, the United States protects citizens from the risks posed by natural disasters through mitigation such as developing fire and building life safety codes, establishing a national system of emergency management, using national hurricane warning programs, and promoting sound land use planning based on known hazards.

In the medical practice, mitigation can involve actions such as buying insurance to cover potential losses, securing shelves to walls, and developing and implementing a plan to reduce susceptibility to hazards. The medical practice should evaluate what types of emergencies are most likely to occur within its geographic area and appropriately prepare.

Staff must be trained on what to do in the event of emergencies, based on the building's structure and geographic locale. Depending upon the types of damage possible, what are the best escape routes for the staff and patients? If those exits are blocked, what alternatives are there?

Many medical practices have windowless areas. For this reason, broad-beamed flashlights should be readily available. Flashlights must be kept in the same place at all times or wall-mounted above the reach of children so in the event of an emergency, everyone on the staff immediately knows where to find a source of light.

If in a tornado-prone or severe storm area, a weather radio should be purchased. It is critical that the practice has at least one battery-operated radio. In the event of a disaster, power lines may be down, telephone communication cut off, or television transmission interrupted. The ability to pick up emergency broadcasting, with up-to-the-minute reports and instructions, is essential. Larger practices should have battery-operated radios in each quadrant of its office.

Cell telephones should be considered as well. If the phone lines are down, cell phones may be the only means of communication with rescue squads or other emergency agencies. However, if there is a natural disaster, hundreds—if not thousands—of other people will also be trying to get through, so cell phone circuits may be busy. Similarly, the cell towers may be down as well.

Consider having a supply of bottled drinking water and nonperishable foods on hand. A disaster could happen during business hours, and the staff and patients might be confined to the area for more than just a few hours. As is true with fire safety, the staff must know how to cooperate with outside emergency agencies. The following information about floods and flash floods, severe storms, earthquakes, tornadoes, and hurricanes can help medical practices handle specific natural disaster situations.

Floods and Flash Floods

Floods are the most common and widespread of all natural disasters. Most communities in the United States can experience some degree of flooding after spring rains, heavy thunderstorms, or winter snow thaws. Most floods develop slowly over a period of days. Flash floods, however, are like walls of water that develop in a matter of minutes. Flash floods can be caused by intense storms or dam failure. Consider the following when preparing for floods:

- Ask the local emergency management office whether the facility is located in a flood plain. Learn the history of flooding in the area. Learn the elevation of the facility in relation to streams, rivers, and dams.
- Learn evacuation routes. Know where to find higher ground in case of a flood.
- Establish warning and evacuation procedures for the facility. Make plans for assisting employees who may need transportation.
- Inspect areas in the facility that are subject to flooding. Identify records and equipment that can be moved to a higher location. Make plans to move records and equipment in case of flood.
- Purchase a National Oceanic and Atmospheric Administration (NOAA) weather radio with a warning alarm tone and battery backup. Listen for flood watches and warnings. Table 14.4 explains the difference between a flood watch and a flood warning.
- Ask the insurance carrier about flood insurance. Regular property and casualty insurance does not cover flooding.

Severe Winter Storms

Severe winter storms bring heavy snow, ice, strong winds, and freezing rain. Winter storms can prevent employees and patients from reaching the facility, leading to a temporary shutdown until roads are cleared. Heavy snow and ice can also cause structural damage and power outages. Following are considerations for preparing for winter storms:

- Listen to NOAA weather radio and local radio and television stations for weather information. Table 14.5 offers explanations of various warnings and advisories.

TABLE 14.4
Flood Alert

Flood Watch	Flooding is possible. Stay tuned to NOAA radio. Be prepared to evacuate. Tune to local radio and television stations for additional information.
Flood Warning	Flooding is already occurring or will occur soon. Take precautions at once. Be prepared to go to higher ground. If advised, evacuate immediately.

Reproduced from Federal Emergency Management Agency, *Emergency Management Guide for Business & Industry*, October 1993.

TABLE 14.5
Severe Winter Storm Alerts

Winter Storm Watch	Severe winter weather is possible.
Winter Storm Warning	Severe winter weather is expected.
Blizzard Warning	Severe winter weather with sustained winds of at least 35 mph is expected.
Traveler's Advisory	Severe winter conditions may make driving difficult or dangerous.

Reproduced from Federal Emergency Management Agency, *Emergency Management Guide for Business & Industry*, October 1993.

- Establish procedures for facility shutdown and early release of employees.
- Store food, water, blankets, battery-powered radios with extra batteries, and other emergency supplies for employees who become stranded at work.
- Provide a backup power source for critical operations.
- Arrange for snow and ice removal from parking lots, walkways, loading docks, and other medical practice areas.

Earthquakes

Earthquakes occur most frequently west of the Rocky Mountains, although historically the most violent earthquakes have occurred in the central United States. Earthquakes usually strike suddenly, violently, and without warning. However, earthquakes are not always sudden. The worst ones begin with a low-rumbling, slow-acting movement. Like the tide, an earthquake is most dangerous as it gains momentum. Generally, the Richter scale reading does not determine damages, but rather how long the quake lasts. For example, a 7.5 quake that lasts for 1 minute will probably not cause as much damage as a 6.0 quake that lasts for 3 minutes.

Earthquakes can seriously damage buildings and their contents; disrupt gas, electric, and telephone services; and trigger landslides, avalanches, flash floods, fires, and huge ocean waves called tsunamis. Aftershocks can occur for weeks following an earthquake.

In many buildings, the greatest danger to people in an earthquake is when equipment and nonstructural elements, such as ceilings, partitions, windows, and lighting fixtures, shake loose. Identifying potential hazards ahead of time and advance planning can reduce the dangers of serious injury or loss of life from an earthquake.

Before an earthquake occurs, check the workplace for hazards such as:

- Shelves securely fastened to walls.
- Large or heavy objects placed on lower shelves.
- Heavy items, such as pictures, hung away from where people sit.
- Overhead light fixtures braced.
- Flammable products securely stored in cabinets with latches and on the bottom shelves.
- Safe places identified inside, such as under sturdy furniture or against an inside wall and away from where glass could shatter around windows, mirrors, pictures, or where heavy bookcases or other heavy furniture could fall over.
- Safe places located outdoors, such as in the open *away* from buildings, trees, or telephone and electrical lines.

Disaster supplies should be purchased in case of an emergency. During an earthquake, attempt the following:

- Ideally before the full brunt of the earthquake, get into a closet, bathroom, or other small space. If not available, take cover under an archway or inside a doorway. Construction that provides additional support is a preferred safety area. For example, if approaching an overpass when a quake begins, pulling over directly under the support columns or walls is safer than being exposed to an unsupported expanse. If needed, abandon the car to seek a safer place till the quake has subsided.

- Take cover under a piece of heavy furniture or against an inside wall and hold on.
- Stay inside. The most dangerous thing to do during the shaking of an earthquake is to try to leave the building because objects can fall on people.
- If outdoors, move into the open, away from buildings, streetlights, and utility wires.
- If in a moving vehicle, stop quickly and stay in the vehicle. Move to a clear area away from buildings, trees, overpasses, or utility wires. Once the shaking has stopped, proceed with caution. Avoid bridges or ramps that might have been damaged by the quake.

After an earthquake, remember to:

- Drive very slowly and cautiously to be sure the road is safe.
- Be prepared for aftershocks, which cause additional damage and may bring weakened structures down. Aftershocks can occur just moments after the full quake, further weakening structures, roads, and so forth. Aftershocks can also occur in the first hours, days, weeks, or even months after the quake.
- Help injured or trapped persons. Give first aid where appropriate. Call for help.
- Listen to a battery-operated radio or television for the latest emergency information.
- Help patients who may require special assistance, such as infants, senior citizens, and people with disabilities.
- Stay out of damaged buildings.
- Use the telephone only for emergency calls.
- Clean up spilled medicines, chemicals, or flammable liquids if it is safe to do so and if the staff have the appropriate equipment to clean up the spilled material. Leave the area if there are gas fumes or fumes from other chemicals.
- Open closet and cupboard doors cautiously to avoid falling contents.
- Have professionals inspect utilities.

Tornadoes

Tornadoes are incredibly violent local storms that extend to the ground with whirling winds that can reach 300 miles per hour and are a treacherous part of the spring season. Tornadoes can occur in any state, but occur more frequently in the Midwest, Southeast, and Southwest. They occur with little or no warning. Each year about 1,000 tornadoes touch down in the United States. Only a small percentage actually strikes occupied buildings, but injuries and fatalities are common. Employees should be educated in the procedures to be taken when a tornado threatens while at work.

In tornado-prone areas, medical practices should develop a severe weather action plan. Planning considerations include the following:

- Ask the local emergency management office about the community's tornado warning system.
- Assemble a disaster supplies kit, including a first-aid kit, canned food with opener, battery-powered radio, flashlight and extra batteries, and written instructions on how to turn off electricity, gas, and water if authorities advise such action.
- Purchase a NOAA weather radio with a warning alarm tone and battery backup. Listen for tornado watches and warnings.
- Establish procedures to inform personnel when tornado warnings are posted.
- Designate tornado shelter areas and inform employees of their location. Areas that provide protection include basements, interior rooms, and hallways on the lowest floor and away from windows. Find a safe room, and either lie flat on the floor (preferably under something) or hug an inside wall while lying on the floor.
- Consider space requirements for shelter areas. Adults require about 6 square feet of space.
- Conduct tornado drills.
- Once in the shelter, personnel should protect their heads with their arms and crouch down.
- Listen to the radio for instructions or for the cancellation of the warning.

The staff should monitor weather information when conditions could potentially produce a tornado. If a tornado warning is issued for the area, an alarm will sound and the office should be evacuated to the designated safety area. Table 14.6 identifies the two types of tornado alerts.

Also be alert to what is happening outside. The following are some observations that people have described when they tell about a tornado experience:

- A sickly greenish or greenish-black color to the sky.
- A strange quiet that occurs within or shortly after the thunderstorm.
- Clouds moving by very fast, especially in a rotating pattern or converging toward one area of the sky.
- A sound a little like a waterfall or rushing air at first, but turning into a roar as it comes closer. The sound of a tornado has been likened to that of both railroad trains and jets.

TABLE 14.6

Tornado Alerts

Tornado Watch	A tornado is likely in the area. Stay tuned to radio and television stations for additional information.
Tornado Warning	A tornado has been sighted in the area or indicated by weather radar. If a tornado warning is issued for the area, an alarm will sound and evacuation should begin in an orderly manner to the designated safety area.

Reproduced from Federal Emergency Management Agency, *Emergency Management Guide for Business & Industry*, October 1993.

- Debris dropping from the sky.
- An obvious funnel-shaped cloud that is rotating or debris, such as branches or leaves, being pulled upward even if no funnel cloud is visible.

If employees report they see a tornado and it is not moving to the right or to the left, relative to trees or power poles in the distance, it may be moving toward them. Although tornadoes usually move from southwest to northeast, they also move toward the east, the southeast, the north, and even northwest.

Hurricanes

Hurricanes are severe tropical storms with sustained winds of 74 miles per hour or greater. Hurricane winds can reach 160 miles per hour and extend inland for hundreds of miles. In the United States, the hurricane death toll has been greatly diminished by timely warnings of approaching storms. The National Weather Service issues hurricane advisories as soon as a hurricane appears to be a threat. Table 14.7 provides a description of a hurricane watch and a hurricane warning. When a hurricane strikes, it is unlikely employees will be caught off guard at work, but damage to fixed property continues to mount. Employees should listen carefully to local authorities to determine what threats can be expected and take the necessary precautions to protect themselves, their families, and their property. In the final analysis, the only real defense against hurricanes is the informed readiness of the people living and working in the community.

When preparing for hurricanes, planning considerations include the following:

- Ask the local emergency management office about community evacuation plans.
- Establish facility shutdown procedures. Establish warning and evacuation procedures. Make plans for assisting employees who may need transportation.
- Make plans for communicating with employees' families before and after a hurricane.
- Purchase a NOAA weather radio with a warning alarm tone and battery backup. Listen for hurricane watches and warnings.
- Survey the facility. Make plans to protect outside equipment and structures.

TABLE 14.7

Hurricane Alerts

Hurricane Watch	A hurricane is possible within 24 to 36 hours. Stay tuned for additional advisories. Tune to local radio and television stations for additional information. An evacuation may be necessary.
Hurricane Warning	A hurricane will hit land within 24 hours. Take precautions at once. If advised, evacuate immediately.

Reproduced from Federal Emergency Management Agency, *Emergency Management Guide for Business & Industry*, October 1993.

- Make plans to protect windows. Permanent storm shutters offer good protection.
- Consider the need for backup systems, such as portable pumps to remove flood water, alternate power sources, or battery-powered emergency lighting.
- Prepare to move records, computers, and other items within the facility to another location.

Technological Emergencies

Technological emergencies include any interruption or loss of a utility service, power source, life support system, information system, or equipment needed to keep the business in operation. Planning considerations include the following:

1. Identify all critical operations, including:
   - Utilities, such as electric power, gas, water, sewer systems, and compressed air
   - Security and alarm systems, life support systems, HVAC systems, and electrical distribution system
   - Manufacturing and pollution control equipment
   - Communication systems, both data and voice computer networks
   - Transportation systems, including air, highway, railroad, and waterway
2. Determine the effect of service disruption.
3. Ensure that key employees are familiar with the building systems.
4. Establish procedures for restoring systems. Determine the need for backup systems.
5. Establish preventive maintenance schedules for all systems and equipment.

Bomb Threats

In no setting is the threat of a bomb a cause for greater concern than in a facility dedicated to the care of people who are ill and infirmed. Bomb threats are acts of terrorism, and emergency planning must include this threat. Action must be taken on the basis of planning and not the distractions and fears of the moment. Coordination is essential. Courage and calm must be the two elements that prevail if another hazard (eg, panic) equally severe to an explosion is to be averted.

Preventive steps can be taken in regard to bomb situations. These steps include the same security safeguards that should be in everyday use to protect the medical office against many other vulnerabilities. For example, locking equipment rooms, utility closets, and storage areas can help reduce the problem by limiting access to these areas. Limited access and reporting suspicious persons and vehicles should be part of the daily protection system and can help minimize vulnerability and the probability of an actual bomb detonation. Additional security measures may also be necessary depending on the type of practice and geographic location.

One of the unique factors of a bomb threat is the guessing game it forces the threatened medical office to play. Is it real or is it a hoax? Do we search? Is evacuation necessary? These basic questions become the framework of the written bomb threat response plan, which must be flexible to the degree that the information received can be applied to these questions at the time of the crisis.

One important step in dealing with a bomb threat is to have the person who receives the call listen carefully and get as much information as possible from and about the caller. Figure 14.1 provides an example of a form to be filled out by the person receiving the call. This form, or a similar one, should be conveniently located at switchboards and other phone locations.

Once the information has been gathered and the telephone call has ended, a supervisor and the appropriate law enforcement authority should be immediately notified. Follow the specific orders of the law enforcement authority, including instructions regarding search and evacuation.

Terminating the threat is an important part of the organizational reaction to the bomb threat. There should be an officially declared ending. All persons who have been notified of the receipt of the threat should be officially informed when the organization is resuming normal operations.

EMERGENCY PLANNING REQUIREMENTS

A variety of government agencies and private organizations provide emergency planning guidelines and requirements, including OSHA, NFPA, FEMA, EPA, the JCAHO, many insurance companies, and state and local governments. Some "authorities having jurisdiction" may require compliance with standards and codes that are different or more stringent from federal standards. NFPA codes and standards form the basis of many OSHA regulations.

Occupational Safety and Health Administration

OSHA emphasizes that employers should establish effective safety and health programs and prepare their employees to handle emergencies before they arise. Their Standard Number 29 CFR 1910.38 requires employee emergency plans and fire prevention plans. The following information summarizes OSHA's Fact Sheet, *Emergency Preparedness and Response*,[6] and outlines the intent of the standard.

Planning
The effectiveness of response during emergencies depends on the amount of planning and training performed. Management must show its support for the importance of emergency planning and employees must be involved. The plan should be comprehensive enough to deal with all types of emergencies specific to the medical practice and consider:

- The type of medical practice (general practitioner or specialist).
- The types of patients that are treated at the office.
- What type of equipment is needed to provide adequate treatment to patients. Depending upon the nature of the practice, some medical practices use more flammables than others or some practices have special equipment or tools (eg, cauterization) that need to be heated and then permitted to cool.

FIGURE 14.1

Sample Response-to-Bomb-Threat Form.
Reproduced with permission from Chaff & Co

RESPONSE TO BOMB THREAT

If you receive a bomb threat, you may be able to reduce the hazards in a real situation.

Instructions:

1. BE CALM; BE COURTEOUS; LISTEN.
2. DO NOT INTERRUPT THE CALLER.

Date _____ Time _____ Person receiving call _____

Exact WORDS of person placing the call: _____
_____

Questions to Ask:

1. When is the bomb going to explode? _____
2. Where is the bomb right now? _____
3. What kind of bomb is it? _____
4. What does it look like? _____
5. Why did you place the bomb? _____

Try to determine the following:

Caller's Identity: ☐ Male ☐ Female ☐ Adult ☐ Juvenile Age ____
Caller's Voice: ☐ Loud ☐ Soft ☐ High ☐ Deep ☐ Raspy
 ☐ Pleasant
Caller's Accent: ☐ Local ☐ Not Local Foreign Region _____
Caller's Speech: ☐ Fast ☐ Slow ☐ Distinct ☐ Disordered
 ☐ Stutter ☐ Slurred ☐ Lisp
Caller's Language: ☐ Excellent ☐ Good ☐ Fair ☐ Poor
 ☐ Foul ☐ Other
Caller's Manner: ☐ Calm ☐ Angry ☐ Rational ☐ Irrational
 ☐ Coherent ☐ Incoherent ☐ Deliberate
 ☐ Emotional ☐ Righteous ☐ Laughing
 ☐ Intoxicated ☐ Drugged
Background Noises: ☐ Office Machines ☐ Factory Machines
 ☐ Bedlam ☐ Train ☐ Music ☐ Quiet
 ☐ Voices ☐ Mixed ☐ Airplanes ☐ Party
 ☐ Street Traffic

Time Caller Hung Up: _____

Additional Information: _____

Action to take immediately after the call:

1. NOTIFY YOUR SUPERVISOR
2. TALK TO NO ONE ABOUT THE CALL OTHER THAN INSTRUCTED BY YOUR SUPERVISOR.
3. BE PREPARED TO REPEAT SAME NOTIFICATION TO POLICE DEPARTMENT, AS REQUESTED.

The plan must include, at a minimum, the following elements:

- Emergency escape procedures and emergency escape route assignments
- Procedures to be followed by employees who remain to perform critical shutdown operations
- Procedures to account for all employees after emergency evacuation is completed
- Rescue and medical duties for assigned employees
- The preferred means for reporting fires and other emergencies
- Names or regular job titles of persons or departments to be contacted for further information or explanation of duties under the plan

The plan should address all potential emergencies that can be expected in the workplace. Therefore, it will be necessary to perform a hazard audit to determine toxic materials in the workplace, possible hazards, and potentially dangerous conditions. For information on chemicals, the manufacturer or supplier can be contacted to obtain MSDSs. These forms describe the hazards that a chemical may present; list precautions to take when handling, storing, or using the substance; and outline emergency and first-aid procedures.

Contingency planning is also important. For example, what backup plans are in place in the event of an emergency? What if the electricity goes out for the entire block? What if the bulk of medical supplies is destroyed?

Chain of Command

A chain of command should be established to minimize confusion so employees will have no doubt about whom has authority for making decisions. A coordinator and a backup coordinator must be designated. Training for this responsibility includes the following:

- Assessing the situation and determining whether an emergency exists that requires activating the emergency procedures
- Directing all efforts in the area, including evacuating personnel and patients
- Ensuring that outside emergency services, such as local fire departments, are called in, when necessary
- Directing the shutdown of operations, when necessary

Communication

During emergencies it is often necessary to evacuate the offices. Also, normal services, such as electricity, water, and telephones, may be nonexistent. Under these conditions, it may be necessary to have an alternate area to which employees can report or that serves as an alternate headquarters.

Alarms must be distinctive and recognizable as a signal to evacuate the work area. The employer must explain to each employee the means for reporting emergencies, such as manual pull-box alarms, public address systems, or telephones. Emergency phone numbers should be posted on or near telephones, on employees' notice boards, or in other conspicuous locations. An updated written list of key personnel should be kept, listed in order of priority.

Under OSHA's Standard (Standard Number 29 CFR 1910.165), the employee alarm system must provide a warning for necessary emergency action as called for in the emergency action plan or provide reaction time for safe escape of employees from the workplace.

Emergency alarm system requirements also include installation, maintenance and testing, provisions for evacuation, emergency reporting procedures, and emergency telephone numbers.

Accounting for Personnel

Management will need to know that all personnel have been accounted for. This can be difficult during shift changes or if contractors are on-site. A responsible person must be appointed to account for personnel and to inform police of those persons believed missing.

Emergency Response Teams

The responsibility for emergency response can be given to either a team of employees or to individuals who have been given specific assignments. If there is a large enough staff, consider putting together an emergency response team. If the practice is small, designate certain individuals who will have specific responsibilities, provide training, and make those tasks part of their job description. In many small medical offices, all staff is trained to be emergency response team members.

Appropriate emergency response is the first line of defense in emergencies. Before assigning personnel to these teams, consideration should be given to the following factors, beyond an understanding of disaster prevention:

- Emotional capacity to remain calm during an emergency
- Ability to maintain an authoritative manner without being dictatorial
- Physical strength to use fire extinguishers or to move heavier equipment out of the way in the event of an emergency
- Superior communication skills to interface with staff, patients, and emergency crews (eg, firefighters, police, paramedics)

In addition, emergency training may include:

- Use of various types of fire extinguishers
- First aid, including CPR
- Shutdown procedures
- Evacuation procedures
- Chemical spill control procedures
- Trauma counseling

Emergency response teams should be trained in the types of possible emergencies and the emergency actions to be performed. They are to be informed about special hazards, such as storage and use of flammable materials, toxic chemicals, radioactive sources, and water-reactive substances, to which they may be exposed during fire and other emergencies. Team members must be able to determine if the fire is too large for them to handle or whether search and emergency rescue procedures should be performed. If there is the possibility of members of the team receiving fatal or incapacitating injuries, they should wait for professional firefighters or emergency response groups.

Training

Training is important to the effectiveness of an emergency plan. Before implementing an emergency action plan, a sufficient number of staff must be trained to assist in the safe and orderly evacuation of employees, patients, and visitors. Training for each type of disaster response is necessary so employees know which actions are appropriate to each situation.

In addition to team member (if there is an emergency response team) training, all employees should be trained in the following:

- Evacuation plans
- Alarm systems
- Reporting procedures for personnel
- Shutdown procedures
- Types of potential emergencies

These training programs must be provided as follows:

- Initially, when the plan is developed
- For all new employees
- When new equipment, materials, or processes are introduced
- When procedures are updated or revised
- When exercises shows that employee performance must be improved
- At least annually

The emergency control procedures should be written in concise terms and be made available to all employees. Drills should be held at random intervals and the group's performance evaluated by management and employees. When possible, drills should include groups supplying outside services, such as fire and police departments. In buildings with several places of employment, the emergency plans should be coordinated with other companies and employees in the building. The emergency plan should be periodically reviewed and updated to maintain adequate response personnel and program efficiency.

Personal Protection Equipment (PPE)

Emergencies should be reported immediately to the proper authorities. Employees and patients should evacuate, if necessary. Effective PPE is essential for any person who may be exposed to potentially hazardous substances. In emergencies, employees may be exposed to a wide variety of hazardous circumstances, including:

- Chemical splashes or contact with toxic materials
- Falling objects or flying particles
- Fires and electrical hazards

Medical Assistance

In a major emergency, time is critical in minimizing injuries. Send injured employees and patients to the hospital.

Security

During an emergency, it is often necessary to secure the area to prevent unauthorized access and to protect vital records and equipment. It may be necessary to notify local law enforcement personnel or to employ private security personnel to secure the area and prevent the entry of unauthorized people.

Records that also may need to be protected include essential accounting files, legal documents, and lists of employees' relatives to be notified in case of emergency. These records may be stored in duplicate outside the medical practice or in protected, secure locations within the facility.

ENDNOTES

1. Ursano RJ, Fullerton CS, Norwood AE. *Psychiatric Dimensions of Disaster: Patient Care, Community Consultation, and Preventive Medicine.* American Psychiatric Association. APA Online: www.psych.org.
2. The United States Fire Administration. *During or After a Disaster Fact Sheets: Flood Fire Safety, Summer Storm Fire Safety, Tornado and Hurricane Fire Safety, Winter Storm Fire Safety.* Office of Fire Management Programs; Emmitsburg, MD. USFA Online: www.usfa.fema.gov.
3. National Fire Protection Association. *NFPA 101 Life Safety Code (2000 edition).* National Fire Protection Association, Inc.
4. National Fire Protection Association. *Sparky's Activity Book.* National Fire Protection Association, Inc, Quincy, MA. Item No. WR-SPY-29C.
5. Federal Emergency Management Agency. *After the Fire! Returning to Normal.* Federal Emergency Management Agency/United States Fire Administration; Emmitsburg, MD. Document Number FA-46/June 1998.
6. OSHA. *Emergency Preparedness and Response Fact Sheet.* Occupational Safety and Health Administration, US Department of Labor; Washington DC.

chapter 15

Education and Training

Training is an important component of a comprehensive safety program. The purpose of training is to educate all employees about their roles and responsibilities in creating, and their rights in expecting, a work environment free from hazards.

Training empowers all employees, supervisors, and physicians. Active participation in the health and safety program by all parties demonstrates commitment to the program and to solving problems when hazards are identified. They can demonstrate this commitment by showing a thorough knowledge of the medical practice's safety program and by understanding the existing and potential hazards to which they may be exposed.

Effective training is the most efficient way to guarantee the success of programs. In addition to disseminating program information, skill instruction heightens employee involvement and labor/management cooperation. Increased knowledge and enhanced human relations have a favorable affect on the program and the facility's image.

SETTING UP A SAFETY TRAINING PROGRAM

Conduct training and education in the language and at a level of understanding appropriate for the individuals being taught. Use evaluations of training effectiveness to revise educational programs.

The planning process is a great opportunity to help employees feel confident that they are working in a safe environment. For example, new employees expect to be shown where things are kept, how tasks are performed, how to use unfamiliar equipment, and so forth. If job training also includes safety training, then safety performance becomes part of an individual's job description and can be included in his or her job performance evaluations.

Safety training for long-time employees involves different approaches. The staff must understand that safety precautions and regulations are not arbitrary standards set up by a manager or a group of doctors. Much of the information is mandated by regulatory agencies, such as OSHA and other organizations at the federal, state, or local levels. While some OSHA guidelines require records that document training, it is also a good idea to keep records for internal use. An annual review of the records can help determine whether training needs to be repeated or refreshed. These regulations exist to protect the employees, patients, and any others who may be on the premises.

A successful, well-run business of any kind welcomes innovative ideas, accepts new challenges, seeks better solutions, finds more effective applications, and puts them to the test. In a manufacturing setting for example, creating a prototype is an accepted business practice and the cost is justified by eliminating the "bugs" or problems before the product is implemented. This practice, which also includes training and education, is appropriate to all industries, including medical practices.

Even safety training must be run like a business. Start by understanding the business issue, not the training need, then select an effective and efficient learning solution that may or may not involve training. Being effective means delivering training services that tangibly help the medical practice achieve its goals. Being efficient means making the true costs of training clearly evident and highly acceptable. (See Table 15.1.)

PLANNING FOR SUCCESSFUL SAFETY TRAINING

Different types of medical group practices (eg, family practice, orthopedic, plastic surgery) have safety training needs that are relevant to their specific practice. This section deals with general safety training that is

TABLE 15.1

Guidelines to Help Make the Costs of Training Clearly Evident and Highly Acceptable

1. *Link safety training objectives to business strategy.* Trainers, whether staff or contract, must understand the practice's business objectives so they can lead participants in the same direction.
2. *Address the corporate culture.* To create long-lasting organizational change, trainers must understand the influence of the culture. This understanding includes learning barriers, such as fear, blame, reluctance to take responsibility, self-justification, and so forth. Corporate culture supports the emotional process of learning.
3. *Focus on outcomes.* Prior to developing a learning initiative, trainers must focus on business results. Be clear about what should be accomplished, what outcomes are wanted, and how the outcomes will be measured. For example, is it to reduce needlesticks or back injuries or to enhance knowledge of new PPE?
4. *De-emphasize formal training.* Traditional training, such as face-to-face classroom instruction, is only one way of teaching. A lot of learning occurs very naturally on the job through team meetings, conversations with supervisors, self-study, or reading industry magazines. Learn about all of the different ways that people can gain knowledge on the job and focus learning activities to those natural inclinations.
5. *Allow employees time to process what they have learned.* In the Information Age, knowledge is not hard to come by, but it takes time to process. The challenge is to find ways to create more time for employees to process new information so their new understanding allows them to draw relevant and creative conclusions.
6. *Demand results from training suppliers.* Many vendors tend to focus heavily on selling their products (ie, safety modules, safety posters, online learning programs) rather than on the results and outcomes of their training programs. Be careful to find a training vendor with the proper mix of business goals and product goals.
7. *Be active in dealing with return on investment.* Calculating return on investment can be made easier by following the aforementioned guidelines. The process becomes clearer when aligning training around business objectives and safety goals.

Reproduced with permission from *Learning Revives Training*, published by ACC Communications, Inc. Copyright 2000. For more information visit www.workforce.com.

appropriate for every medical practice including selected safety training considerations for specialty practices.

General Safety Training

To provide a successful safety training program, the medical practice must: establish training needs, select the training application, confirm planning responsibilities, design the training program, conduct instructor-led training, evaluate the training, and anticipate additional training needs.

Establish Training Needs

There are certain areas that apply to most medical practices when identifying employee safety training. These areas include:

- Infectious disease hazards
- Fire prevention
- Emergency preparedness
- Housekeeping practices
- Patient safety
- Hazardous chemical exposures
- Ergonomics
- Personal safety
- Recordkeeping

Not all problems or situations require training as a solution. To be successful, training must have a goal. When responding to problems (eg, incidents) or specific situations (eg, such as new technology) a needs analysis helps identify the goal (ie, what training to conduct). Finding out who needs the information and why simplifies this process. For example, do employees need to know about new regulations? Do certain employees need new job skills? Have injuries or illnesses occurred that require refresher training? Are there too many mistakes being made? Are there new employees who need information specific to their jobs?

Take into consideration the individual characteristics of the staff. Having established where safety training is necessary, determine how to provide it. Such demographics as average age, educational background, location of medical practice, previous experience, and language proficiency all have an influence on learning.

Select the Training Application

Once training needs are identified, determine how to meet those needs. In-house staff may do some training. In other cases, outside trainers may be necessary. For example, equipment manufacturers provide training in the safe and correct use of their equipment.

If there is a hospital in the area, ask its safety director to address the staff. If the physicians in the practice belong to any special organizations, determine if there are training materials or consultants who could provide insightful information. Government agencies, such as OSHA, NIOSH, CDC, and EPA, have considerable free material to assist in training.

Technology-based learning is another option becoming widely used as medical facilities face greater challenges retaining knowledgeable employees. Self-directed learning, such as interactive CDs, distance learning, strategic classroom learning, and virtual universities, are examples of technology-based learning.

Confirm Planning Responsibilities
Someone with organizational skills must coordinate the various aspects of a successful training program. If the training is instructor-led, is the seating comfortable and arranged in such a way that participants can be easily heard (eg, questions, input)? Is the room or area free from extraneous noises? Is the temperature comfortable?

If the training is by interactive CD or Internet-based, structure is important. There should be quizzes throughout a successful course. Employees should work through case studies where they get a chance to practice what they learned. Finally, there should be exams. In some cases, computer-based training must be followed with an instructor-led class so employees can ask questions or participate in demonstrations.

Design the Training Program
Designing the training program includes deciding how to meet various needs (eg, team activities, audiovisuals, study guides, interactive CD) and how to evaluate if the needs have been met (eg, practice with coaching, testing, observing, feedback).

Employees should be asked to complete course evaluations at the conclusion of the program. (See Figure 15.1 for a sample course evaluation.) Were the program objectives met? Did the training meet the needs of the participants? What did participants learn? Was the room comfortable? Was the trainer knowledgeable and enthusiastic? The answers to these questions and others form the basis of action for improving the next training session.

Conduct Instructor-Led Training
To build camaraderie with trainees, use various participation-based techniques. Participation creates an opportunity for people to actually take part in their own learning experience, which translates first into increased knowledge of the subject and ultimately into productivity on the job. Just as important, activities give people a chance to succeed at something during the training process.

These approaches are different from earlier training styles where employees were shown or told what needed to be known (ie, with little or no solicitation of questions) or given a manual to memorize. If employees understand *why* a safety procedure or regulation exists, they are more likely to comply.

Evaluate the Training
Evaluating the effectiveness of training is essential to safety programs and a fundamental aspect of adult instruction. Evaluation is a continuous process of collecting and interpreting information. It is a dynamic ongoing activity that contributes to the refinement of the teaching–learning process.

FIGURE 15.1

Sample Course Evaluation Form

Reproduced from *How To Manage Training: A Guide to Adminstration, Design, and Delivery.* Copyright 1991 Carolyn Nilson. Used by permission of the publisher, AMACOM, a division of American Management Association International, New York, NY. All rights reserved. www.amanet.org.

Evaluation Form
COURSE EVALUATION FORM (Trainee)

How to Use This Form
1. Distribute forms to all trainees in the course. Do this at a logical break in the instruction—about two hours before the end of the course.
2. Tell trainees that you appreciate their thoughtful responses in their role as evaluator. Thank each one as the completed form is turned in to you.
3. Make it clear that putting their names on the form is optional. What you really want are their opinions about the quality of the course, including specific suggestions for improvement.
4. Allow adequate time—at least twenty minutes. Stop class a little early if necessary to get a careful response. Encourage "additional comments."

Instructions:
Place an X in the column 1,2,3,4 to represent your evaluation of each item.

Negative   Positive
1   2   3   4

Learning Objectives
 1. Appropriate, learnable, pitched right
 2. Organized to facilitate learning
 3. Clearly stated
 4. Exercises helped accomplish objectives

Content
 5. Accurate
 6. Current
 7. Adequate in scope
 8. Sequenced properly

Course Setting
 9. Comfortable
 10. Quality materials and visual aids
 11. Adequate equipment

Instructor's Delivery
 12. Course agenda and timing of lessons
 13. Engaging presentation style
 14. Respectful of trainee contributions
 15. Preparation and expertise

Relevance to the job
 16. Course content relevance
 17. Relevance of instructional techniques
 18. New skills useable right away

(continues)

FIGURE 15.1
continued

19. New skills will save time
20. New skills will save money
21. Your confidence level to use this training on your job

Additional comments and suggestions for improvement:

_____
_____
_____
_____

Would you recommend this course to others? Who? Why?

_____
_____
_____

Evaluation is ongoing, not just something that is done at the end of the program. Part of the evaluation process determines if the objectives have been met and the employee's skill level has improved. Trainers should show the effect of training on the bottom line. This means that trainers should focus on outputs, which are the business results of training. According to the 1999 *State of the Industry Report* prepared by the American Society for Training and Development, a scant 15% of training courses are evaluated based on business results.[1]

Instead, the most popular evaluation method continues to be the reaction of participants. However, just because someone liked a training course does not mean it enhanced that person's performance back on the job. Senior management focuses on business needs and outcomes, and trainers should focus on the same things.

Furthermore, to properly evaluate a training session, there must be involvement with the medical practice's core mission. The training should support the practice's objectives. There are several ways to evaluate the effectiveness of a training session. The following list includes tools that are used by educators to help in the evaluation process:

- *Checklist* evaluates performance based on specific criteria and as a guide that indicates whether or not a sequence of behavior has been achieved.
- *Pretest*, used at the beginning of a training session, determines if the employee has already met the objectives and is ready for additional work.
- *Post-training quiz*, used at the end of a training session, determines what the employees have learned in the session. Quizzes can be graded and then suggested reading or study information can be instantly provided to the individual based on information that was missed.
- *Direct observation* evaluates actual practice or performance after the training session.
- *Review of problems or situations* that led to the training session evaluates training. For example, incident reports and statistics on employee injuries may have determined the need for education on safety or body mechanics. Incident reports reviewed after the training can determine if the course had a positive effect in reducing injuries.

Feedback is an important element in the evaluation process. There is a tendency to tell the participants when they are wrong, but to say nothing when they are right. Feedback reinforces positive attitudes and redirects negative behavior. It is important for employees to know if they perform a task well. It offers a stronger possibility that they will continue to correctly perform the task when the situations arise. Feedback is also used to provide information to the trainer. An individual may be asked for an appraisal of a session. A group may be asked for written appraisals upon conclusion of a unit of instruction or major segment. Conference sessions may be held in order to provide an informal, relaxed climate.

Ongoing interaction and reinforcement stimulates commitment among employees. Short briefings, minitraining programs, or updates with newsletters or incentive programs can all contribute to emphasizing management's commitment to safety and health and motivating employees' cooperation.

Employees are valuable resources for improving training results. By utilizing post-session evaluation forms, employees can suggest what additional training is needed, if any.

Anticipate Additional Training Needs

Training vigilance is also essential in a medical practice because of constant advances in technology and methodology. In recent years, safety training has experienced a new focus and direction. For example, as more is learned about the effects of chemical substances on human beings, employees must be trained or retrained on exposure limits and protection precautions. As chemical additives are altered, so are the potential effects upon humans. NIOSH is mandated to collect, evaluate, and analyze scientific and medical research and recommend methods of chemical exposure control for workers. OSHA promulgates standards to protect workers from hazardous chemicals. The EPA regulates the environmentally safe disposal of hazardous chemicals. Employers and employees have a responsibility to work safely by taking ample precautions when using hazardous chemicals. Employee training and retraining are major components of the process.

Other trends, such as consumer awareness, have increased the need for well-organized, up-to-date safety educational programs for employees.

Selected Training Considerations for Specialty Practices

This section highlights some of the selected training considerations for specialty practices.

Orthopedic and Occupational Therapy Tools and Saws

Orthopedic and occupational therapy saws, sanders, driers, drills, and hand tools (ie, pliers, screwdrivers, chisels, snips) can be sources of injury to employees and patients. In addition, the dusts created from various operations while using these tools can be a major source of indoor air quality problems, as can the fumes from various glues and solvents used in creating and/or fitting orthopedic devices or operating occupational therapy practices.

The OSHA 29 CFR1910 standards address the requirements for training employees in the safe use of powered and hand tools.

In general, safety training in the use of powered and hand tools should include the following information:

- Shop safety and housekeeping.
- Storage of tools and chemicals.
- Safety checks of power tools to include frayed electrical cords and switches that do not correctly function.
- Safety checks of hand tools to include cracks, breaks, loose heads, and pitted blades.
- Electrical safety checks.
- The use of machine guards.
- The use of fume hoods.

- The use of a dust-collecting system.
- The use of PPE.

Radiology and Mammography Equipment

Many medical practices have X-ray machines. Some specialty and general practices also have mammography equipment. Safety training is necessary, and special rooms, warnings, and monitoring must be in place to protect the staff and patients. Compliance with applicable local, state, and federal guidelines and standards concerning the hazards surrounding radiological equipment must also be addressed both in the safety program and in the training program. The OSHA Standard (Standard Number 29 CFR 1910.1096, *Ionizing Radiation*) addresses OSHA's requirements for staff involved with the use of or possible exposure to ionizing radiation.

The safety training program should include staff who have had specialized training in the use and operation of the radiological equipment and staff who will be working around radiological equipment, but not operating it.

In general, training for the safe use of radiological equipment should include at least the following information:

- Familiarization with terms that are related to dangers associated with radiological equipment.
- Design and safeguards associated with rooms that house radiological equipment.
- Types of, and reasons for, personal monitoring for radiation exposure.
- Familiarization with the type of, and reason for, annual certification of radiological equipment and shielding.
- Familiarization with policies and procedures that address the use of warning signs, warning lights, damage to shielded areas, electrical safety of the equipment, shooting the X ray, and the processing of radiological films.
- Proper body mechanics to use when transferring and assisting patients onto and off of X-ray tables.

Defibrillators

Staff members who are involved in the operation and maintenance of defibrillators must be specially trained in their use, calibration, and safety checks needed to prevent serious staff and patient injuries. If a defibrillator causes a serious injury to staff or a patient, the incident must be reported under the SMDA.

In general, safety training for using a defibrillator should include at least the following information:

- How the defibrillator operates.
- How to store and charge the defibrillator.
- How to test the defibrillator to ensure that the voltage being asked for is actually the voltage that the machine is outputting.
- Electrical safety of the switches, cords, and plugs.
- How to handle an electrocution.
- The designation of competent persons to operate and maintain the defibrillators.

Sterilizers

Sterilizers and autoclaves are often used in medical practices. This equipment is used to sterilize surgical instruments and laboratory specimens; however, the same autoclave used to sterilize equipment should not be used to sterilize laboratory biological specimens.

There are many safety issues with sterilizers, including equipment failure, exposure to live steam and very hot items, weakening and explosion of the pressure vessel, and incorrect use of the equipment leading to nonsterile objects. Equipment manufacturers are usually available to provide training sessions for the operation and maintenance of this equipment.

Safety training for using a steam sterilizer or autoclave should include at least the following information:

- The theory of operation of a steam sterilizer.
- How the particular brand of sterilizer operates and the type of maintenance required.
- How to use biological indicators to ensure that the objects are sterilized.
- What PPE to use to protect the staff from burns.
- Where to set the sterilizer up to minimize the possibility of inadvertent staff or patient contact with this hot piece of equipment or the freshly processed and hot items.
- The protocols, policies, and procedures designating competent staff that will be operating the steam sterilizer.

FUNDAMENTALS OF SAFETY TRAINING

There are certain fundamentals of safety training that are relevant to every medical practice. They include the following:

1. An introduction to the current safety culture at the facility. This process covers management's commitment to the employee, the employee's responsibilities for working safely, what resources are available, and who to contact in event of an emergency (eg, injury, fire).
2. Adherence to regulatory agency requirements that are applicable to employees' jobs and what is necessary to comply.
3. Accident prevention techniques, such as how to correctly lift items or patients and the importance of cleaning up spills (ie, from coffee to water to chemicals).
4. The status of the facility's emergency preparedness.
5. Fire prevention training, including location of extinguishers and how to operate them and evacuation procedures for patients and staff.
6. How the practice ensures patient and staff safety, including housekeeping practices and using medical equipment.
7. The value of PPE and how it protects the employees.
8. Techniques to protect employees from work-related disorders, including back injuries and repetitive stress disorders.
9. How and why to report an accident or incident and how the information is used.

10. The practice's philosophy on workers' compensation.
11. Workplace violence prevention.

Additional program elements for certain medical practices include the following:

1. Infection control guidelines.
2. Requirements of the Bloodborne Pathogens Standard.
3. Occupational exposure to tuberculosis.
4. Preventing airborne infections.
5. Hazardous chemical training.
6. How to dispose of regulated medical waste.
7. Working with hazardous materials, such as cytotoxic drugs.
8. Reducing indoor air pollution.

REGULATORY AGENCY TRAINING REQUIREMENTS

Many of the elements listed in the previous section include the training that is required by regulatory agencies. Some standards spell out how often employees are trained, including what the training must cover. In certain situations, OSHA may require that employers provide employees with training and education programs in addition to relevant job descriptions or information, including guidelines for:

- Safe working conditions and precautions.
- Health and safety precautions against possible hazards on the job.
- How to detect symptoms of exposure to substances in the workplace.
- Emergency treatment procedures.

TRAINING TECHNIQUES

In the race to remain competitive in today's marketplace, medical practices need employees who are highly skilled, productive, and able to work together as a team. In an effort to develop these types of workers, medical practices sometimes come up against two major drawbacks. First, in many cases, training budgets are low. Second, employees to be trained frequently are accustomed to receiving information in a fast-paced, entertaining manner. Because of this, they become disinterested during yet another "boring" video or classroom-style lecture.

Interactive training can increase the value of training without inflating the budget. This approach concentrates on boosting the interactivity and the "fun content" of training. Employees learn and retain more because they are involved in the training and because the experience is pleasurable. Another benefit to interactive training is that it makes employees partners in the learning process. While the manager or supervisor may be responsible for communicating information, the employee is responsible for getting involved in activities, asking questions, completing exercises, and sharing ideas with coworkers. In many cases, the training curriculum can be found in the day-to-day business of the medical practice. By focusing on the problems that the practice is experiencing and asking for employees' input, a trainer can provide vital learning opportunities to the trainees as well as introduce possible solutions to the problems.

Make sure that the safety training activities have meaning for the work environment. Humor or entertainment just for its *own* sake does not work. Do not be afraid to experiment. Music, humorous incidents, and stories are all tools of the trade. Games and puzzles also stimulate creativity and imagination during training. New ideas can come alive that bring value to many areas of the practice, including streamlining safety procedures or implementing unique patient appreciation services. Creativity helps to develop confidence and skills that enhance work and everyday life skills.

Additional creative training approaches (see Table 15.2) reinforce learning and communication. The skills that the employees learn can be designed to improve their people-interaction ability. Safety training sessions can include:

- Handling difficult situations.
- Improving listening skills.
- Learning how to go the extra step to solve a problem.
- Understanding how our own behavior affects the way others see us.
- Increasing communication among departments.

Training techniques chosen should not be too conservative, but, at the same time, they need to target the audience and the objectives of the medical practice. The appropriate use of inventive training strategies not only increases the volume of what is learned and later practiced, but also makes training a much more enjoyable experience for everyone involved—even new trainers. (See Table 15.3 for guidelines on selecting teaching methods.)

TABLE 15.2

Creative Approaches to Learning

- *Music.* By the proper use of music in training, learning takes on a new dimension. Music affects emotions and motivation.
- *Reflections.* Self-image is important. Training helps employees understand that what they think about themselves is as important as what others think about them.
- *Listening.* Learning how to solve employee or patient problems requires listening to what they are saying and why it is important to them. Many times what someone says is not exactly what he or she means. By teaching good listening skills, employees are able to empathize with others.
- *Choosing emotions.* Employees can be placed in training situations that teach them about emotional reactions. They learn to solve problems effectively without becoming emotionally involved. Training can show them how to find and implement the most effective solutions.
- *Surprises.* Unexpected situations can occur in training that require employees to practice the skills they are learning. These situations simulate surprises that can occur when they are working. Someone who is prepared handles situations better than someone who has never seen the situation before.
- *All-the-time skills.* Employees appreciate acquiring skills that they can use at home as well as at work. Behavior cannot be turned on and off depending on the surrounding environment—work or home. Learning how to utilize service skills in everyday life helps make the skills an integral part of employees' personalities.

Reproduced with permission from *Safety Guide for Health Care Institutions, Fifth Edition* written by Linda F Chaff published by Health Forum, Inc. Copyright 1994. All rights reserved.

Other training techniques, such as the following, can be effective, depending on the goal of the session(s):

- *Discussion.* With discussion, the instructor is not so much in charge, but is serving as a host or hostess to a group of people. The instructor acts as a facilitator, someone who moves the subject along or presents questions not brought up by the other group members. Specific problems are usually the topics for the meeting. The objective is for the group to interact in a participatory learning atmosphere, presenting concepts, discussing problems, and coming to solutions in a creative and cooperative manner. In general, a discussion session has a primary purpose and involves key personnel.

- *Group brainstorming.* This technique includes verbal interaction between members of a group, whether a problem needs to be resolved, services improved, or needlestick incidents reduced. Ideas, no matter how farfetched or irrelevant, are spoken. No one's words are meant to be a final decision, but merely springboards to alternative possibilities. A highly successful solution may be derived from one casual suggestion simply because others saw the possibilities from a different perspective. Table 15.4[2] presents some tips on how to get the most out of a brainstorming session.

- *Demonstration or skills training.* This is extremely useful for tasks that require manual dexterity or several steps to complete, such as training on how to operate a piece of equipment. If the equipment is new to the facility, the instructor introduces what it is, what it's meant to accomplish, and how the task is to be safely performed, step by step. Trainees should be positioned, whether seated or standing, with a clear view of the procedure.

TABLE 15.3

Guidelines for Selecting a Teaching Method

- Determine facility resources—staff, funds, time.
- Decide which options will fit within budget restrictions.
- Determine the appropriate method for the audience.
- Match the technique to training needs. For example, skills training requires hands-on learning, but explaining a new regulation may only require lecture and discussion.
- Apply new teaching methods gradually and with explanation, especially for groups accustomed to lecture only.
- Choose only those techniques with which you, the trainer, are comfortable. If you feel awkward or inadequate using a specific method, avoid it until your experience level allows you to feel more comfortable with the strategy. When you are having difficulty, the students will perceive it and may take your instruction less seriously.
- Use only the audiovisual materials that relate to the topic at hand. Avoid the use of films or videotapes as entertainment only. Provide ample time before and after any audiovisual aid to discuss it, and help the audience integrate the material. The audiovisual aid should seldom, if ever, be the focus of the entire training period.
- Preview all audiovisual materials and read all written training materials before discussing them with the class.
- Learn from mistakes and bad choices of training methods, using student evaluation forms.

Reproduced with permission from *Safety Guide for Health Care Institutions, Fifth Edition* written by Linda F Chaff published by Health Forum, Inc. Copyright 1994. All rights reserved.

If possible, provide a flipchart or chalkboard with the process already written out so students can follow each step, both visually and in a written format. The more complex the task, the more important it is to have oral, visual, and written instructions as part of the training session.

Once demonstrated, students then perform the task themselves. The instructor supervises this hands-on training until satisfied that one or more trainees have mastered the safe operation of the equipment. Having accomplished that, the supervisor may either wish to spend more time with those trainees who have not quite learned each step in sequence or perhaps assign the trainees to another trainee who has mastered the process.

- *Role-playing.* Role-play can be an effective method for exploring how to cope with situations and people and effectively deal with them. Another advantage to role-playing is it can be a means of measuring previous training. Much of the response from trainees is molded by how the concept of the role-play is presented.

    Role-plays can be used to demonstrate to employees the difference between safe and unsafe work practices. This process is most effective when done by professional trainers or actors so employees benefit from the dynamics of the role-play, instead of trying to correctly do the role-play.

    By asking questions and inviting discussion, role-plays can become a vital, enlightening, and entertaining way of learning. Discussion not only helps trainees learn about their own jobs, but also to better understand other employees' responsibilities.

- *Case study.* This approach consists of reviewing a realistic situation and responding to it. From a minor incident to a worst-case scenario, case studies are an effective means of training because they focus on specifics, not broad generalizations.

- *Traditional self-study.* This method involves preprogrammed lessons and exercises that the trainees may study individually at home or elsewhere. Self-study may involve a text or manual of some kind, often with review questions and answer keys.

TABLE 15.4

Tips on How to Get the Most Out of a Brainstorming Session

- *Stay loose.* People generate more ideas when they have fun. They feel free to say anything and to spin ideas off each other.
- *Don't judge.* Negative criticism brings the free flow of ideas to a screeching halt. New ideas are like delicate bubbles. They can be popped by a sneer, yawn, quip, or frown.
- *Write down everything.* Choose a group scribe whose job it is to capture all the team's ideas. Write on large sheets of paper. Cover the walls with the sheets so everyone can see every idea.
- *Be outrageous.* Have fun. Be off the wall. Go for outlandish ideas. There will be plenty of time to sort through and evaluate the ideas later.
- *Evaluate.* When it is time to select specific ideas or narrow the group's focus on the best ideas, have participants select their favorite three ideas on the sheets of paper. Very quickly there will be a sense of the most popular ideas.

Adapted from *Five Star Mind*, published by Doubleday Canada Limited. Copyright 1995.

The value of self-study is that it permits trainees to progress at their own pace. However, its value depends entirely upon the students' willingness to learn without supervision or interaction with others. Not everyone reacts well to self-study, so it is a good idea to review some of the assigned material early on. In that manner, the level of independent motivation can be established before too much time has elapsed. Self-study must have provisions for interaction with a knowledgeable person and site-specific issues. Staff who do not understand English very well or who are not comfortable reading, may not do well with this type of training.

An observant trainer quickly determines what type or types of training would be best for the majority of the staff. Trainers who involve participants and keep them interested deliver high value to organizations.

Creative safety training approaches reinforce learning and communication. The skills employees master can be designed to improve people-interaction ability in working safely among departments.

TRAINING AND LEARNING EVOLUTIONS

As medical practices continue to merge or partner with other health care facilities, additional safety training options become available. Many health care facilities and organizations are making dramatic changes in the way they develop the safety knowledge and skills of their workforce with technology-based learning. Self-directed learning, such as interactive CDs, distance learning, strategic classroom learning, and virtual universities, are examples of technology-based learning. Forces that have been reshaping the marketplace for the last two decades are bringing about the transformation. Table 15.5 presents the most influential of these change agents.

TABLE 15.5

Influential Change Agents in Self-Directed Learning

- *Global competition.* This pressure eliminates the luxury of long program development cycles. Employees must learn as needed, not when it is convenient.
- *Right-sizing and other euphemisms.* Competition and other pressures continue to force medical practices of every size to cut costs. Mergers and acquisitions continue to redefine the corporate landscape. The employees who survive these reorganizations often find themselves saddled with more responsibility, yet lack the knowledge, experience, and skills to do what is expected of them. Providing relevant, effective, and cutting-edge training helps the employee meet the challenge.
- *Changing employer–employee relationships.* The entire employer–employee relationship has dramatically changed. Employees know that the responsibility for their careers lies in their ability to stay abreast of critical knowledge and skills; employers realize that they must provide these opportunities to more of their people. Learning is being put back in the hands of employees.
- *Technology.* No other development has had a greater effect on the way we do business than the ever-increasing use of and accessibility to high-technology tools. Technology is having a profound affect on training.

Reproduced with permission from *Goodbye Training, Hello Learning*, published by ACC Communications, Inc. Copyright 1999. For more information visit www.workforce.com.

Some self-directed learning packages, however, do not meet the intent of OSHA's training requirement that safety training be specific for the organization and that a competent person be available to answer questions before the employee begins work. Before purchasing programs, they should be evaluated for compliance and flexibility regarding this OSHA requirement.

Based on current trends, self-directed learning, distance learning, strategic classrooms, and corporate "virtual" universities are gaining popularity as medical practices continue to develop the knowledge and skills of their workforce.

Self-directed Learning

Interactive computer-based training can serve as a valuable training tool in the context of an overall training program. However, use of computer-based training by itself is not sufficient to meet the intent of most of OSHA's training requirements. This technique includes computer or Internet-based training as a form of self-study. This concept is gaining popularity because employees can often train at home or at the office. No matter how many training courses the practice offers, there are times when knowledge is needed, but a facilitator or trainer is unavailable. Trainers can later review results and progress and be available for help or additional information. This type of training is a tool and is unlikely to completely take the place of traditional training using instructors or facilitators.

Some busy employees use independent learning via the World Wide Web to meet job requirements or to satisfy interest in a particular subject. Employees are not tied to a classroom and many times courses are available for registration any time of the year. Participation requires a computer with an Internet connection. Independent learning via the Internet requires motivation and self-discipline. Some employees find that this type of learning provides an opportunity to more effectively mix course work with family and career responsibilities.

When Web-based training is provided at the medical practice, consider having the computers in an area away from the work desk. People are accustomed to learning in a classroom and they like to go someplace other than their desks or cubicles.

Self-directed learning offers a number of advantages, including the availability of the class in several languages. Travel time and expenses are reduced. Because the programs are already developed, expensive programming costs are eliminated. Productivity is maintained, even enhanced, because learners can take the training on their own time. After the initial equipment and software purchases, online training is less expensive than many instructor-led courses. There is a level of consistency that cannot be reached with traditional, instructor-led training, especially for medical practices with multilingual workforces. When self-directed learning is blended with instructor-led classes, participants come to class "warmed-up" at relatively the same level, reducing training time.

The disadvantages of such training include costs. This type of training can be a costly investment in terms of equipment and software. In addition, many people have concerns about reducing group interaction and the ability to build on ideas. If a medical practice relies too heavily on self-directed learning, employees may miss out on such opportunities as meeting and networking with colleagues and managers. When this type of training method is used, the medical practice needs to consider how to facilitate interaction and communication in different ways.

Distance Learning

Distance learning, such as through interactive videoconferencing network capabilities, offers an unparalleled opportunity to provide high-quality education and access not otherwise possible. Compressed digital interactive video is a telecommunications technology that provides instruction through electronic media to students and individuals not physically present at the site of the class. Unlike satellite communications, which is one-way, compressed digital interactive video is a two-way live television system. Participants can see, hear, and speak to individuals at each site connected for a class or conference.

Strategic Classrooms

There will always be the need for the interpersonal, face-to-face contact only available in a classroom setting. However, because the demands and needs of today's workforce are shifting so dramatically, classroom training must be adapted to complement the easy access and on-demand capabilities of technology-based learning. When technology and the classroom work in concert, the learning process can be much more efficient and targeted.

When this strategic combination is used, technology-based learning enables participants to acquire the knowledge and practice the skills before the students step into a classroom. A well-designed technology course charts progress, provides feedback, and ensures that completion of the program guarantees a certain level of competency. Not only is classroom time used more efficiently, but the content of the program can be more effectively targeted as well. Content can also be linked more closely to specific safety objectives.

Corporate "Virtual" Universities

In some cases, there is literally no brick-and-mortar facility, only a virtual facility that may have a production studio. These learning centers are developed to quickly change to meet the new demands of technology. Successful virtual universities keep their curriculums fluid, constantly adding new courses and new methodologies. Corporate universities are usually adopted by large corporations or health care organizations.

LEVELS OF TRAINING

Different positions in the medical practice require distinct training approaches. The four levels of training that should be treated separately are:

1. New employee orientation
2. Current employee training

3. Performance improvement and upgrading of long-term employees
4. Supervisor training

Within each of these areas, a needs assessment process is essential. The major purpose of the process is to identify and then meet the ever-changing needs of the individual, department, and medical practice. Therefore, determining needs is an ongoing process for the educator/manager. Assessing needs serves several purposes including:

- Identifying the beginning point for instruction
- Focusing the material to be covered
- Answering the question, "Why should the content be taught?"
- Giving permission to stop at a given point and conclude when information has been learned

New Employee Orientation

The newcomer's first impression of an organization is important. If the hiring process produces increasingly qualified and well-suited new workers, orientation reinforces an already existing sense of safety in the facility. The hiring process alone constitutes an investment, and it is important that the investment is utilized to the best advantage.

The new employee has much to learn about the organization. Although this does represent a great deal of work, freshness is also an opportunity. The new employee has not become entrenched in routines that overlook safety. New employees can be welcomed and gently indoctrinated with the safety philosophy of the facility.

Safety orientation should be broad and include an introduction to each of the minimum elements of safety. To avoid information overload, safety training of new employees should be spread out over a number of days. In this way, groundwork is laid not only for safe practices early in the employee's term of service, but also for later safety training. Employers and coworkers have a legal and a moral responsibility for the safety of new workers. Some health care facilities practice what has come to be known as a *training marathon*—where numerous subjects are quickly covered in a short amount of time and without much detail. This practice is exhausting for both the trainers and the new employees and presents a casual and potentially dangerous approach to workplace safety. Management should not minimize the necessity of orientation. Well-trained employees are more productive, happier, and more quickly come up to standard.

The first step toward safety for a new employee comes when the employer designs the job and answers questions such as:

- What training is required?
- Who delivers it?
- How is training paced?

The second important step in protecting the new employee is adequate supervision. Supervisors should spend time making sure new employees understand the safety reasons why certain tasks must be performed in specific ways. With some more specialized jobs, it may be written into the design of the new employee's job descriptions of how often supervisors check on their work.

This initial training period is an ideal time for senior management to demonstrate its support for the safety program and for the new employee who will practice it. This can be done within the first week of employment, which is a good way to instill in new employees the enthusiasm that the program needs. After this probationary training period, the employee should continue to receive instruction with other employees through in-services or seminars.

Current Employee Training

Nonmanagement education should focus on developing skills that are learned through repetition. To this end, policies and regulations should be translated into procedures that employees can practice. Repetition ensures that safe actions become routine. These procedures are more likely to be carried out if attention is paid to practicality. To be sure procedures are practical, training sessions need to rely heavily on give-and-take in which employees' reactions and ideas are solicited and used. This participation both strengthens the program and enhances learning by building confidence and enthusiasm.

Another strategy, which increases what employees actually take back with them to their jobs as they leave an in-service training session, is experiential learning. Simply put, this means that the employees "do" rather than just "listen." They must also listen as the instructor discusses theory, standards, policies, and procedures, but not in place of training time spent in actual practice.

Although practical methods are the primary thrust in training for nonmanagement personnel, the importance of theory should not be overlooked. Employees are more likely to resent a new program when no one has bothered to explain its purpose. Accordingly, the introduction of the program to staff should begin with the reasons why it is necessary. Instructors should emphasize that escalating risks not only endanger the staff, but even threaten the facility's existence as well. At the same time, employees should know that throughout the course of implementation, their feedback and views would continue to be important to management.

In this way, the entire staff learns ways to comply with new safety regulations and to get along with coworkers and managers. The combination of skills and communication are then the heart of the safety program, as it becomes a natural component of the practice's routine.

Long-Term Employee Training

It is equally essential to the safety program that long-term employees receive regular instruction. The elements to be included change somewhat when training employees who have held their jobs for 2, 5, 10, or 20 years. One aspect that changes is instruction on the techniques of the job. Generally, if employees have held their jobs for any length of time, they know what to do. It should not be necessary to go through the lengthy process of discussing each step of their job. However, there are certain areas on which to focus the training for long-term employees. These areas are as follows:

- New techniques that have developed for minimizing the risks of specific areas, such as infectious materials or sharps
- New regulations and requirements
- Recent phenomena, such as HIV and tuberculosis
- New safety policies and procedures

First, new techniques constantly develop for dealing with safety-related situations. These techniques may also make a job easier or less time-consuming. The employees require instruction to develop these new skills.

Second, jobs change with time. Restructuring often occurs, making a job title that meant one thing yesterday mean something rather different today. When this takes place, the practice usually upgrades long-term employees with training. Furthermore, with ever-progressing technology, employees should be instructed in how to use new equipment and how to carry out new forms of treatment.

Another subject for training of long-term employees is new safety policies and procedures. Effective safety programs are constantly modified to meet new regulations, changing technology, and the needs of the medical practice. As a result, policies and procedures are frequently revised or rewritten. Employees must understand these changes in policy and know how to perform the revised safety procedures.

When conducting this type of in-service training, differentiate long-term employees from new employees. Explain that the reason for this training is to introduce the employee to new aspects of the job and to explain changing requirements. In addition, remember to give the reasons for changes and new policies. This eases long-term employees' anxiety about change, facilitates a clear understanding of why it is important, and helps them develop a more positive attitude about the message.

Supervisor Training

Today's supervisor must demonstrate successful leadership characteristics that include energy, good judgment, human relations' skills, and technical competence. A few may be born leaders, but most people are good supervisors primarily because they have been trained and have worked hard to become so.

In addition to these traits, two characteristics crucial to supervisory ability are good communication and people skills. Most supervisors would like to have friendly and productive interactions with their subordinates because ways in which supervisors communicate with their employees have a profound affect on the success of programs at the medical practice. Interactions establish the climate within each company, and this climate motivates employees toward better performance and risk avoidance on their jobs.

Formal training and development can be excellent mechanisms by which managers cultivate the abilities necessary for leadership. During this education, supervisors must learn safety policies and the ways to carry out procedures, just as all other personnel. Supervisors should also be taught how to explain the regulations to their employees and how to detect unsafe acts. This step should not be overlooked because support from supervisors increases employee commitment and involvement. Armed with in-depth knowledge of the program and its specific areas, the supervisor can ensure that the safety program evolves into an everyday component of the facility's activities.

Training that covers program goals, policies, and procedures should be appropriate, with recent and accurate information. In addition, it should stress management's vital role in developing and implementing the system of safety management. This prepares supervisors to be representatives of program philosophy and practical application methods.

Generally, supervisory training should:

- Explain the topic at hand, such as "Right-to-Know Laws," and how this program fits into the facility's larger comprehensive safety system.
- Show how the specific program will be put into place. Approaches are important in gaining the commitment necessary for success.
- Provide skills training in the specific areas of the program as they relate to each department (eg, proper lifting techniques training).
- Build support for the safety program by communicating desired management behaviors.
- Comply with applicable regulations, such as OSHA's Standard for Bloodborne Pathogens.

Beyond these functions, management should also receive instruction in developing good leadership skills. Many consulting services offer supervisory training that helps managers become good leaders. Preferably, management skills training should be woven in with the other training topics. Thus, supervisors simultaneously become both better leaders and more knowledgeable managers.

GENERATION-SPECIFIC TRAINING

The dynamics of the American workforce is changing to include both an older population of workers as well as young workers. Medical practices must keep pace with this new workforce by offering varied training.

The Aging Workforce

The American workforce is growing older. The average age of employees continues to rise with the aging of the generation known as the baby boomers—those born between the years 1945 and 1964. The Hudson Institute's *Workforce 2020* reports that by 2020, nearly 20% of the US population will be 65 or older. This means that the number of Americans of retirement age will equal the number of Americans aged 20 to 35.[3] Figure 15.2 and Figure 15.3 provide statistics on the growing labor force.

Retirement does not interest many older Americans. As they enjoy longer and healthier lives, aging baby boomers are changing their minds about retiring at 65. According to the American Association of Retired Persons, 80% of today's baby boomers plan to continue working at least part-time after age 65, as compared to the 22% of 65- to 69-year-olds working today.[4]

Research indicates that job performance increases with age, particularly in the areas of accuracy and consistency. In addition, when a person's

FIGURE 15.2

Female Labor Force Participation Rates

Reproduced with permission from *Workforce 2020*, published by Hudson Institute. Copyright 1997. For more information visit www.hudson.com.

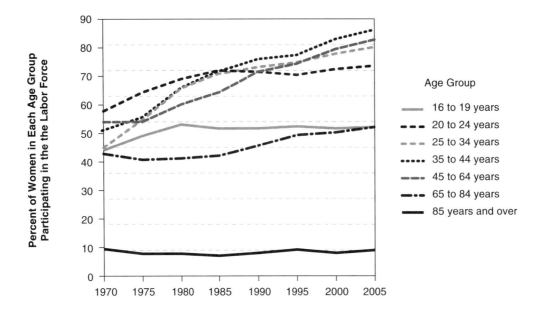

FIGURE 15.3

Male Labor Force Participation Rates

Reproduced with permission from *Workforce 2020*, published by Hudson Institute. Copyright 1997. For more information visit www.hudson.com.

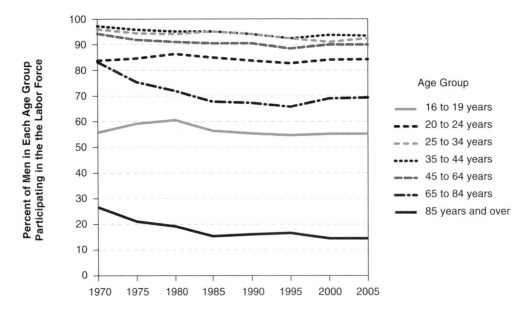

abilities are matched to job requirements, good job performance usually results, regardless of age. Seniors are usually eager to upgrade their skills and ready to do whatever the job takes. They are very loyal, take less time off, are customarily on time to work, and are less likely to get injured at work.

Although good job performance and a strong work ethic may be a plus for older workers, safety concerns remain. Safety professionals are challenged to identify needs and develop programs to keep the aging worker safe, healthy, and productive. Changes may occur in the aging workers' vision, hearing, heart, lungs, skin, muscles, urinary tract, intestinal tract, and immune system.

Falls are often an issue in workplaces. Inadequate lighting, loose carpeting or flooring, steps, or even file cabinet drawers that are left open contribute to falls. In many cases, older workers' reflexes slow down and eye–hand coordination weakens. In addition, eyesight and hearing may become diminished, along with their sense of balance. All of these changes can contribute to falls, as well as other types of injuries.

It is important to look at ways to train older workers so they function effectively in their jobs and make the most of their experience and skills. Retraining of certain tasks is also important, such as lifting and moving equipment or materials. For example, older workers may be trained to get help lifting an object that weighs more than 30 pounds or retrained on how to properly lift heavy objects to reduce back strain.

Repetitive tasks, such as inputting data into a computer, are more common in today's workplace. These tasks can lead to back problems, carpal tunnel syndrome, eyestrain, headaches, low morale, and other health-related problems, regardless of age. The best way to prevent repetitive-task injuries is with ergonomics training and increased awareness.

Rather than forcing the human body to adapt to machinery and equipment, ergonomics allows workers to adjust equipment to fit the body. Proactive programs should be developed to address the special needs of older workers. Table 15.6 provides a variety of ergonomic and other training solutions for the aging workforce.

The Young Workers

The number of young workers in our society has increased during the past few years, while many older workers take early retirement. This can leave a pared down workforce without experienced workers to pass on their knowledge and young workers in need of a steadying knowledgeable hand. Thus, training sessions geared to their specific needs can be a real challenge to management and first-line supervisors.

Understanding Young Workers

Effectively using a young worker's talents depends in large part on how well supervisors understand the unique characteristics of a younger generation that has grown up in an entertainment environment accustomed to quick interactions and rapid video input from television and the movies. The following includes some of these needs of a young worker.[5]

Need for Structure

Young workers have a real need for structure in their lives, especially on the job. They want guidance and supervision, but often are afraid to ask for it. It is essential that they know their job duties. It is the responsibility of the supervisor to spell out clearly the young workers' responsibilities. Young workers must know where they stand.

TABLE 15.6

Ergonomic Solutions to the Changes that Come with Aging

Normal Changes with Aging	Ergonomic Recommendations
Visual Decreased ability to read fine print Decreased adaptation to the dark Increased sensitivity to glare Altered depth perception Reduction in ability of eyes to tear	**Visual** Brighter lighting Reduce/eliminate glare with indirect lighting Use special-purpose lighting Uniform/appropriate contrast materials Visual cues with training
Hearing Decreased high-frequency hearing Decreased ability to discriminate some sounds	**Hearing** Avoid high-frequency noise Reduce background noise Provide equipment with adjustable noise levels
Heart Numerous changes including decreased muscle tone and elasticity Decreased tolerance to cardiac stress	**Heart** Avoid work in extreme temperatures Provide frequent rest breaks
Respiratory Decreased cough mechanism effectiveness Decreased functional reserve capacity (normal for regular activities, but may have difficulty on exertion) Decreased lung expansion	**Respiratory** Avoid work with potential for respiratory irritation/sensitization Evaluate ability to wear respirator in stressful conditions (ie, confined space) Avoid work with excessive exertion (ie, frequent stair or ladder climbing)
Skin Decreased fat and water in subcutaneous tissue Decreased skin elasticity Decreased size and number of sweat glands, more difficulty in regulating body temperature	**Skin** Avoid work in extreme hot or cold temperatures Monitor ability to wear skin barriers Avoid work with chemicals with defatting properties
Muscles Decreased muscle mass and strength Increased muscle response time and fatigue	**Muscles** Avoid or reduce work with static muscle effort Increase use of mechanical lifts Keep work in "lifting zone" Reduce/eliminate twists Encourage stretching and exercise programs
Urinary Decreased bladder capacity Increased urinary frequency Decreased ability to concentrate urine Increased prostrate size	**Urinary** Provide frequent bathroom breaks Provide work with accessible bathroom facilities
Intestinal Slower digestion of food Decrease in liver enzyme concentration Decreased insulin release	**Intestinal** Allow "snack breaks" Avoid work with chemicals with potential toxic effects on liver Alter work schedule to increase time for meals
Endocrine Decreased insulin production Decreased thyroid function Decreased tolerance to heat or cold	**Endocrine** Allow rest breaks Avoid work in hot or cold environments
Immune Decreased inflammatory response Increased risk of infections	**Immune** Special precautions to avoid infection Avoid repetitive-motion work Design job to prevent cumulative trauma injuries

Reproduced with permission from *SafetyFocus Newsletter, November-December 2000, The Baby Boomers are Coming: Is Your Ergonomic Program Ready?* Copyright 2000.

Need for Approval
The need for approval is a dominant force in most young workers who may be insecure in their first job, as opposed to an older worker with years of experience to build confidence. If they perform their job in an acceptable manner, their supervisor should tell them so. Even a smile with a nod of approval can be used effectively by a supervisor to bring out the best in a young worker.

Need for Acceptance
An insecure young worker has a real need to be accepted by those persons considered to be important in life. If the supervisor treats the individual as a valuable person in their daily interaction, the young worker may well decide to work toward gaining recognition and acceptance from the supervisor.

Need for Growth and Development
Most young people today are looking for purpose in their lives and many hope to find it in their work. They want to make not only a contribution to the medical practice, but also to their self-development. The supervisor who recognizes this need for growth and development in the young worker and who allows the young worker to participate in making job decisions, where possible, not only performs a service to the worker, but also contributes to the practice's morale and service.

Factors in Motivation
The question of what motivates a person to perform at a high level on the job has been studied at length over the past few years. A recent study involving 500 young workers in South Carolina revealed the following rank-order of motivational factors that they felt were important in their jobs[6]:

1. Duty to do one's best
2. Personal satisfaction
3. Working conditions
4. Steady work
5. Liking the boss
6. Fair boss
7. Proving that one can do the job
8. A fair company
9. Praise from boss
10. Chance for promotion
11. Chance for raise
12. Pay
13. Peer respect
14. Praise from coworkers
15. Fear of getting fired
16. Fear of reprimand

As is true with most people, if the motivations of self-determination, self-expression, and a sense of personal worth can be tapped, the individual can be more effectively motivated.

The use of external sanctions and threats of pressuring for production may work to some degree, but not to the extent that the more internalized motives do. Young people are more likely to respond more intensely to positive encouragement versus negative threats.

Rights and Responsibilities

As with all employees, the medical practice has the right to require "a full day's work" from its young employees. It has the right to require that its young employees be productive members of the medical practice workforce. A medical practice also has the responsibility to provide proper training and quality supervision to its young workers. This really is what the young worker is seeking.

Safety Concerns Regarding Young Workers

According to the Industrial Accident Prevention Association, there is a direct relationship between job experience and injuries. The top five causes of injury to young workers are[7]:

1. Slips and falls
2. Overexertion
3. Struck by, or against, an object
4. Bodily reaction (toxic effects from chemicals)
5. Burns

Young workers do not have experience to recognize many hazards in the workplace. They need training, awareness, and experience to recognize them. New workers should be taught that some hazards could cause an immediate injury (ie, slipping and falling on a wet floor, eyes splashed by a chemical like phenol, and being burned). New workers should also be taught that other hazards can cause them to become sick or injured over a period of time that they may not notice right away (ie, exposure to hazardous chemicals, regularly lifting heavy boxes, continuous work on computers without frequent eye breaks, poor indoor air quality). Chronic injuries sometimes are not noticed for years.

Sometimes young workers do not tell a supervisor about an injury because they might be afraid the boss will think they cannot properly do the job if an injury is reported, they may think the injury is not significant, or they may be concerned about what coworkers will think.

Young people tend to view older coworkers as much too cautious. This attitude accounts for shortcuts taken by young workers who may feel a bold new approach to a certain job makes safety precautions somewhat unnecessary.

For these and other reasons, new young workers need special safety training to ease them into the workplace. The first step toward safety for a young worker is when the job is designed. Ask questions, such as "What training is required?" "Who will deliver it?" "How will training be paced?" As with more experienced employees, safety training should be spread out over a number of days to avoid information overload.

The second important step in protecting young workers is adequate supervision. Supervisors should spend time making sure young workers understand the safety reasons why certain tasks must be performed in specific ways. It should be written into the design of the young worker's job description about how often supervisors would check on their work.

When assessing jobs to identify possible pitfalls for young workers, ask the following:

- Are there occasional hazards, like having to take trash to a compactor or dealing with equipment malfunctions?
- Will the young worker be in hazardous areas or work with hazardous chemicals?
- Will situations arise where young workers could go unsupervised for long periods of time?

Recognizing job hazards may lead to redesigning of some tasks to remove inappropriate elements and make them more suitable for young workers until they are more experienced.

The tasks employers deem suitable for new young workers are best determined by answering two simple questions:

- Can a new, young employee handle this safely?
- If so, what training is required?

Vague job descriptions may invite accidents when eager, young workers feel an obligation to "help out" with unplanned tasks for which they have received no safety training.

SAFETY TRAINING FOR MULTICULTURAL EMPLOYEES

Immigration trends tell us to expect 820,000 immigrants to arrive annually in the United States. Two out of three will be of working age upon their arrival. By 2050, it is expected that immigration will have increased the US population by 80 million people. Fully two thirds of the projected US population increase will be due to immigration.[8]

Employing international workers is crucial in meeting today's business demands. The following are a few reasons why medical practices seek employees from other cultures[9]:

1. When a medical practice has a multilingual staff, it can be easier to communicate with patients and vendors of various cultures and languages.
2. It can be good customer relations (ie, patients sometimes prefer interacting with a multicultural staff).
3. It deepens the pool of potential hires.
4. It can make the workplace more interesting.
5. Foreign-born workers can fill job needs not met by American workers.

A concern of supervisors of foreign-born workers is English proficiency. An employee may know enough of the language to get the job, but may not completely understand directions. Understanding is especially critical when it comes to safety regulations.

Providing effective safety training for foreign-born workers with English as their second language can create confusion and concern for trainers. Training should include communicating the safety requirements, assessing how well the individual has understood what was said, and understanding what the employees are saying when they are talking or asking questions. Even foreign-born people whose native language is English may use the language differently.

Ways to communicate include picture posters and other visuals, procedures written in the employee's language, and interactive CDs or other self-directed learning tools in the language of the employee. When possible, instructions should be given verbally because employees understand oral instructions more easily than struggling through written ones. To be totally effective, the employee should be taught the concepts of safety in their own language, *first*, and then be taught the same concepts in English because the employee will be hearing, seeing, and communicating primarily in English in the medical practice. Table 15.7 provides some tips for communicating with people during safety training for whom English is a second language.

TABLE 15.7

Tips for Communicating with People During Safety Training for Whom English Is a Second Language

- Don't shout
- Speak slowly and distinctly
- Emphasize key words
- Allow pauses
- Let the employee read the lips of the trainer
- Use visual aids
- Use handouts in the language of the employee
- Be aware of voice tone
- Use as many familiar words as possible
- Repeat and recap frequently
- Take care not to patronize
- Check often for understanding
- Do not cover too much information at one time
- Be careful when translating

Assessing how well the individual understood:
- Watch for nonverbal signs that indicate confusion or embarrassment
- Notice a lack of interruptions
- Notice the complete absence of questions
- Notice inappropriate laughter
- Invite questions in private and in writing
- Be alert to the yes that means "Yes, I hear your question" not "Yes, I understand"
- Be alert to a positive response to a negative question
- Have the listener repeat what was said
- Observe behavior of employees and inspect work

Understanding people who are learning English:
- Share the responsibility for poor communication
- Observe body language
- Remember to listen
- Read the employee's lips
- Give the employee plenty of time to communicate

Reproduced with permission from *How Managers Can Communicate Better with Foreign-born Employees*, published by ACC Communications, Inc. Copyright 1993. For more information visit www.workforce.com.

Trainers must be confident the individuals understand requirements and hazards and they must make sure they are safe members of the workforce.

AUDITING TRAINING RETENTION

Even the best training programs suffer from a drop in retention of course material over time. The greatest decrease occurs when the training material is not frequently used. For example, the OSHA "Right-to-Know Laws" require that employees know what to do in the event of a hazardous chemical spill. Because spills do not occur often, the employee may forget how to respond and require retraining. One of the ways to determine when to retrain is to ask questions that measure the degree of retention. This can be done during regular work times as well as during safety inspections. The results can be documented and fed into a quality control system that triggers retraining.

ENDNOTES

1. Caudron S. *Learning Revives Training.* Workforce, January 2000, Vol. 79, No. 1, pp. 34-37. Available at: www.workforce.com.
2. Wujec T. *Five Star Mind: Games & Puzzles to Stimulate Your Creativity & Imagination.* New York: Bantam Doubleday Dell Publishing Group, Inc; 1995.
3. Judy RW, D'Amico C. *Workforce 2020: Work and Workers in the 21st Century* (p. 5). Indianapolis: Hudson Institute, Inc; 1997.
4. Krispin K. *Working into the Golden Years.* National Safety Council: Itasca, IL. Safety & Health, November, 2000.
5. Safety Info. *Supervisor Training 3: The Young Worker.* Available at: www.safetyinfo.com.
6. Safety Info. *Supervisor Training 3: The Young Worker.* Available at: www.safetyinfo.com.
7. IAPA. *Young Worker Awareness Program.* 1996. Industrial Accident Prevention Association, Toronto, Canada. Available at: www.iapa.on.ca.
8. US Department of Labor. *Futurework: Trends and Challenges for Work in the 21st Century.* Available at: www.dol.gov. January 6, 2001.
9. Raphael T. *Dear Workforce: Why Recruit From Other Cultures?* ACC Communications, Inc. Available at: www.workforce.com.

chapter 16

The Impact of Accidents and Illnesses

Employees should be able to arrive at work, work in a safe environment, and return home in the same condition.

Preventing work-related accidents and illnesses is critical for an effective safety culture. Health care workers can be exposed to a variety of occupational hazards, including repeated physical or emotional trauma, toxic chemicals, awkward lifting situations, and a broad range of infectious agents.

Alternative work arrangements, such as job sharing, part-time scheduling, working longer shifts, and temporary work, are responses to rapid technological and economic changes. These changes present new challenges to assuring the safety and health of employees who work in medical practices. Seemingly unsuspecting circumstances, like fatigue or constantly rushed work environments, can also play a role in workplace accidents.

The National Safety Council (NSC) estimates that in 1999, work-related deaths and injuries cost the nation at least $122.6 billion. This figure includes $63.9 billion in wage and productivity losses, $19.9 billion in medical expenses, and $23.5 billion in administrative expenses. Further, NSC reports in *Injury Facts* that the economic impact of fatal and nonfatal unintentional injuries (ie, at work and off the job) amounted to $469.0 billion in 1999. By building an awareness and understanding of statistics such as these, medical practice employers can make a positive impact on the safety culture of their environment.[1]

One component of a safety program's success depends on the ability of the employer to quantify and track its accidents and illnesses. This process requires a support system where people feel comfortable looking at an accident, talking about it, and reviewing how it occurred. The first step toward providing a safe and healthy environment is recognizing the situations that increase the opportunity for injury or illness, such as the following:

- Do employees report improperly working equipment?
- Do employees (including supervisors) promptly and properly clean up spills?
- Does the individual using a sharp take responsibility for proper disposal?

- Is the work environment often chaotic for physicians, employees, and patients?
- Are employee complaints listened to and acted upon?
- Does management show sincere appreciation for employee input?
- Do employees feel comfortable asking questions or repeating instructions to be sure they understand?
- Is safety a priority to both management and employees?

NATIONAL INJURY AND ILLNESS REPORTS

The Bureau of Labor Statistics (BLS), US Department of Labor (DOL), helps employers quantify and track injuries and illnesses.[2] The BLS reports annually on the number of days-away-from-work injuries and illnesses in private industry and the rate of such incidents. The number and frequency of these cases are based on logs and other records that are kept by employers throughout the year (ie, recordkeeping requirements discussed later in this chapter). Government agencies, researchers, manufacturers, and consultants glean information from these reports to summarize lessons learned, to prepare bulletins and reports to create awareness among industries, and to invent new control mechanisms to prevent recurring injuries and illnesses. One recent benefit of the BLS report is the confirmation of work-related musculoskeletal disorders (WMSD) in the health care industry and exposure characteristics, which help identify the condition resulting from the lost worktime case and how the event or exposure occurred.

The BLS report also documents other trends and characteristics. Worker characteristics detail the demographic of the injured or ill worker by providing not only the occupation but also the gender and age of the worker, occupational group, length of service with the employer at the time of the incident, and race or ethnic origin.

In 1999, the BLS reported there were 1.7 million occupational injuries and illnesses involving time away from work. Sprains and strains are by far the most frequent disabling conditions, accounting for more than 4 out of 10 cases in 1999, most often involving the back. Bodily reaction and exertion (ie, contact with objects and equipment) and falls were the most frequent events or exposures that lead to work injury or illness involving days away from work.

In 1999, 81,600 of the 1.7 million cases with days away from work were classified as illnesses. Illnesses are often more difficult to link with work than injuries. Some conditions (eg, tuberculosis [TB], cancers, central nervous system disorders) are difficult to relate to the workplace and are not adequately recognized and reported. Work-related aspects of illness may go unrecognized for many reasons, including long latency periods between the exposure and the development of some diseases. In contrast, the overwhelming majority of the reported new illnesses are those easier to directly relate to workplace activity (eg, contact dermatitis, carpal tunnel syndrome).

Infections in Health Care Workers

The NIOSH *Worker Health Chartbook 2000*[3] reports that the 10 million health care workers in the United States constitute approximately 8% of the workforce. Surveillance data on infections in these workers appear in four federal health databases:

- National Surveillance System for Hospital Health Care Workers (NaSH) tracks exposures to and infections from several agents, including TB, vaccine-preventable diseases, and bloodborne pathogens.
- The Viral Hepatitis Surveillance Program (VHSP) and the Sentinel Counties Study of Acute Viral Hepatitis track hepatitis infections.
- Cases of AIDS and HIV infection among health care workers are ascertained from several sources, including the HIV/AIDS Reporting System (HARS), which is maintained by CDC.
- Health department TB control programs use *StaffTRAK-TB* to monitor skin testing in employees of their clinics and affiliated institutions.

Consequences of Bloodborne Exposures

The following are consequences of bloodborne exposures[4]:

- *Hepatitis B Virus.* The VHSP and the Sentinel Counties Study of Acute Viral Hepatitis indicate a 93% decline in hepatitis B viral infections in health care workers over a 10-year period. This decline may be attributed to the adoption of universal precautions against exposure to body fluids and vaccinations against hepatitis B.
- *Hepatitis C Virus.* Many physicians and nurses have increased occupational risk to hepatitis C from needlestick injuries.
- *Human Immunodeficiency Virus.* Most documented cases of occupational HIV transmission occurred among nurses (42%) and laboratory workers (35%).
- *Tuberculosis (TB).* Health care workers have long been at risk of contracting TB. This risk increased in the 1980s with the resurgence of TB in the United States and the subsequent development of drug resistant TB bacteria during the AIDS epidemic.

REPORTING INJURIES AND ILLNESSES

Federal and state laws require employers to report specific work-related injuries and illnesses. Insurance carriers also require incident, injury, and illness reports. A comprehensive reporting process includes many elements, such as a method to ensure that employees understand what to report, the importance of reporting information, the data that should be reported, how reports are used, and regular feedback regarding results of reporting.

Employees must be trained in incident-reporting techniques, including when to file a report, how to fill it out, and to whom to send it. During training, the benefits of incident reporting should be emphasized to employees, such as how the reports are used to reduce hazards and how they can prevent future accidents.

What Should Be Reported

Report all incidents whether or not injury or property damage resulted. Incidents should be reported in a standard format and a copy furnished to the medical practice's insurance company. Many facilities use the same format for employee and patient incidents. This form is usually sent to the insurance company before follow-up is complete. Therefore, follow-up information should be documented on a separate form and filed internally once the investigation process is completed.

The policy of some facilities is to report only severe injuries or accidents. This approach, however, is a mistake because less serious incidents, or close calls, often predict the occurrence of more damaging events. Close calls are accidents that did not happen, but could have except for good luck. The purpose of reporting is to:

- Establish how the incident occurred (its causes or hazards).
- Provide documentation of the measures the employer takes to protect employees, patients, and visitors.
- Provide a uniform method for investigating incidents to determine how to keep the situation from happening again.
- Provide statistical information for analysis to determine if the incident is part of a trend.
- Identify management systems that need to be improved.
- Comply with OSHA recordkeeping and insurance reporting requirements.
- Use as a tool for designing employee training programs.
- Increase safety awareness and employee participation in the safety program.

Incident report documentation may also be necessary as evidence in several situations, such as workers' compensation procedures, state and local agency inspections, insurance reviews, and OSHA inspections.

Who Should Report an Incident

Employees involved in an incident or accident should participate in completing the report. When employees are directly involved with providing information, it gives them an insight into the situation, avoids inaccuracies due to relaying the incident through others, and helps encourage the employee to contribute to a safer workplace. Having a supervisor fill out the incident report may give the appearance of a disciplinary action and the employee may become reluctant to tell the whole story.

As employees participate more fully in safety activities, they begin to embrace the safety program's goals as their own. Supervisors should be involved by reviewing the incident report and discussing it with the employee.

What Information Needs To Be Reported

Certain information helps identify whether a trend exists. Some examples of information to report include:

- Why the incident occurred?
- Did any injuries, illnesses, or property damage occur as a result of the incident?

- Where did the incident happen?
- When did the incident occur?

One of the most effective means of establishing corrective procedures is to consult with the employees who routinely perform the tasks. Inviting the employees' input serves several purposes. It provides the most relevant information for improvement, demonstrates management's commitment to safety, encourages the employee to want to work safely, and serves to reinforce the employee's importance to the medical practice.

How to Route the Report

Limit the number of persons within the medical practice who see the report. Distributing the report to numerous individuals violates confidentiality, discourages future reporting, and may fragment follow-up efforts. Give the completed incident report to the individual in charge of safety in the workplace. In a small practice, this may be the bookkeeper, a nurse, or a physician. In a large practice, it might be the office manager or a consulting safety professional. This individual uses the report to follow up on the incident, quantify and track injuries and illnesses, redesign training and procedures if necessary, and comply with OSHA recordkeeping requirements and the reporting requirements of the insurance company.

INVESTIGATING INCIDENTS

When an incident occurs, it should be investigated by the medical practice.

When to Start the Investigation

Investigate injuries, illnesses, and other incidents as soon as possible after they occur. If an employee is injured, start the investigation immediately after the employee has received needed medical assistance or the situation has been stabilized. Delays can easily result in a change in the physical situation and surrounding circumstances. Lighting levels change, temperatures change, liquids evaporate, and, in general, the causative situation changes. Additionally, the recall ability of witnesses to the event changes. The initial investigation should take place at the scene of the incident. If the investigation or interviews must be delayed, take steps to preserve the situation if necessary, such as securing the area, taking photographs, and talking with individuals. Figure 16.1 provides a flowchart of an action plan for responding to injuries and illnesses.

Techniques for Investigation

Those directly involved are usually able to provide the most accurate answers to what happened. However, many things can affect what and how people remember. A serious accident can result in shock that alters one's memory. A minor incident may result in only partially paying attention, so details may be overlooked. Witnesses can be beneficial during investigations. There are two types of investigations: informal and formal.

FIGURE 16.1

A Flowchart of the Action Plan for Response to Employee Injuries

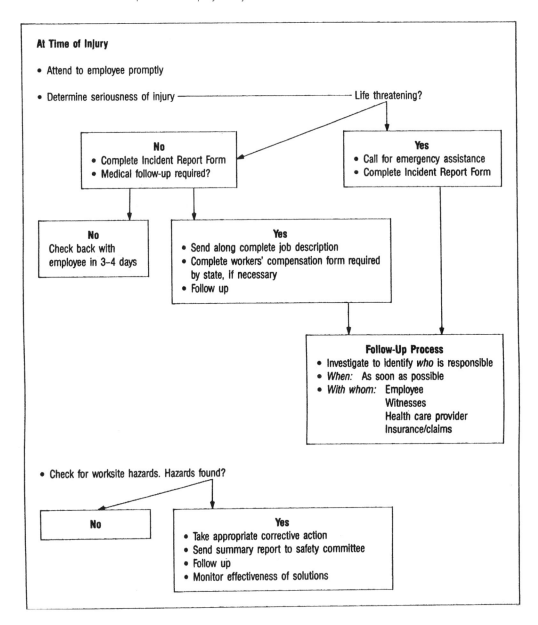

Informal Investigation

An informal investigation is primarily a walk-through with the employee. This is often a general observation of the scene and a short discussion with the employee (or others present) to establish what happened. In the instance of an accident that did not result in injury, review the incident to determine what might have happened and what could still happen if steps are not taken to correct the problem. If necessary, the supervisor might take notes during the walk-through.

Formal Investigation

A formal investigation is often considerably more complex. Generally, it is an in-depth analysis and report of what took place. It would be easy to assume that only incidents resulting in bodily harm, illness, or property damage would require a formal investigation, but that is not necessarily the case. There can also be situations that, if not corrected, could subsequently lead to grave consequences.

Overall, situations or conditions that indicate the need for a formal investigation include:

- An evolving trend (eg, occurrence, locale, people) involved in incidents.
- A series of incidents involving a particular procedure or piece of equipment.
- An increase in similar types of incidents.
- Potentially serious errors in medication.
- Falls (ie, with or without injury).
- A list of similar patient, visitor, or employee complaints.
- Incidents that resulted in hospitalization or fatality.

If several employees complain about eyestrain, the cause is not necessarily poor eyesight. Perhaps there is insufficient lighting to perform a task. Or it could be that staff members are working in a windowless room seated at computers with overhead fluorescent fixtures as the only source of light. Adding incandescent lamps helps to reduce eyestrain without having to move the staff to another area.

Back strain is a common problem in medical practices. Do employees experiencing back strain have a history of back problems? Are employees using proper lifting techniques? Has it been established that the items to be lifted are not heavier than the average person can manage? How can the problem be resolved?

There may be a series of incidents that are caused by *apparently* unconnected sources yet, in combination, produce an unwanted result. At first glance, they might be classified as isolated sources, such as:

- Personal considerations (eg, fatigue, distraction, stress, physical condition, poor balance, poor eyesight, poor coordination).
- Training (ie, inadequate training).
- Staffing levels (eg, insufficient staff, high turnover).
- Supervision (eg, lack of supervision, management deficiencies).
- Unsafe behaviors (eg, ignoring procedures, carelessness, not wearing PPE).
- Unsafe conditions (eg, poor housekeeping, haphazard storage of medical supplies, improper management of chemical hazards).
- Culture (eg, poor morale, autocratic environment, complaints not listened to or solved, high absenteeism).

Each of these factors interacts, to one degree or another, to create a domino effect. For example, a staff member's personal problem might lead to lack of concentration.

As with an informal investigation, time plays an important role in obtaining accurate information. Establish the causes of an incident as quickly as possible. Postponing an investigation until the next day, or longer, can lead to a dimmed memory of the circumstances that led up to the incident or to an unintentional distortion of the facts.

Employing Effective Interviewing Skills

When employees are interviewed following an incident, they may be uncomfortable. This is another reason for conducting an investigation as quickly as possible. Not only is the information still fresh in the minds of employees, but it avoids needless nervousness as well. A good interviewer accepts that the employee or witnesses may be uncomfortable because of unpleasant interview experiences in the past, fear of some form of reprisal, or personal problems. It is necessary to clarify that the investigation is not a blame or fault-finding mission, rather a way to find out how the incident happened and how to avert a recurrence that observation only might not otherwise detect.

Where an interview takes place can also have an influence on the outcome. If possible, conduct the interview at the employee's worksite. If an employee is having difficulty verbalizing what occurred, ask him or her to act it out. Showing what happened can often produce more effective results than statements alone. If an employee states that someone left the room, that is one image. However, when acting it out, if an employee turns left instead of right, it might have a significant effect on how the incident occurred.

A skilled interviewer listens for a witness's tone of voice, which indicates emotional upset or concern. If the witness shows shortness of breath or speaks in clipped, staccato sentences, these indicate nervousness or even apprehension. Body language is another helpful tool. An uncooperative employee—for any reason—might sit with arms folded across his or her chest. Any bodily signs of pulling back or shutting off can indicate the need to get through to the individual to enlist his or her aid. By showing appropriate sympathy, an interviewer can establish a stronger rapport. For example, the interviewer might say, "I realize this has to be very upsetting," or "Take your time, there's no rush."

At the end of the interview, it is very important to thank employees for their time and assistance and end the interview on a positive note. If steps are in progress to avoid a repeat incident, communicate this to the employee. The objective is to indicate that management is committed to a safe and healthy environment for its employees and patients and is doing everything possible to ensure their well-being.

Effective interviewing skills are acquired through training and experience. Table 16.1 provides steps, in acronym form, to help with a successful interview.

TABLE 16.1
"A MESS" Acronym That Spells Out the Steps to a Successful Interview

Acknowledge – Listen to people
Mirror – Rephrase what they say
Empathize – Understand their emotions
Solve – Determine correct action
Settle – Provide satisfactory solution

Reproduced with permission from *People Power*, published by Chaff & Co. Copyright 1992.

Establishing Goals for Improvement

One of the most practical means of establishing goals that are workable and effective is to involve the entire staff. Those who perform the tasks are the most likely to be aware of any need for improved safety. Information gathering is a basic step toward implementing a viable incident-reporting program.

Other than direct observation of employees' task-specific behaviors, a good starting place is the records of incidents already on file. How many needlesticks have there been this month compared to last month? How many back-related injuries—minor or major—have there been compared to last year? If specific patterns of behavior have led to incident reports, the process of identifying areas for improvement becomes clear.

It is important to any incident-reporting goal that the program be ongoing. There may be changes in procedures, practice location, personnel, or any number of factors that can affect the program, causing it to be routinely updated or reassessed.

ELIMINATING ACCIDENTS THROUGH DATA-TRACKING PROGRAMS

Employees may routinely use a comprehensive incident-reporting program that generates extensive amounts of useful information, but simply having the information provides no benefits. The success of incident reporting hinges on what is *done* with the information. Incident reports have many uses, but additional information is needed for the system to meet its full potential.

Summary data from a number of incidents can uncover conditions that are difficult to detect, even by monitoring the individual incidents. For example, incidents that happen months apart and in different departments may form part of a year-long pattern that can be detected only with the recordkeeping and computation that is provided by an organized system of data tracking. Data-tracking systems compile information on numerous variables involved in incidents, such as incident types, time of day, shifts, and job responsibility. In this way, trends and related factors become evident.

A data-tracking system does not need to be elaborate. A simple log (see Figure 16.2) can be used to track and compare many variables. This type of form has space to note what corrective actions were taken and when, providing the ability to monitor the status of action taken on an incident. The form can be effectively completed on the computer or by hand. A good data-tracking system allows the person who is coordinating incident reporting to track and follow up on incidents and to ensure that action is taken and corrective measures are evaluated for effectiveness. Software programs are also available that allow detailed analysis and streamlined recordkeeping.

WORKERS' COMPENSATION

Workers' compensation protects employees who are injured or contract a disease on the job. Workers' compensation insurance in the United States originated as a turn-of-the-century reform of the insurance system designed to maximize benefits to workers, while minimizing administrative and litigation costs.

FIGURE 16.2
Data-Tracking Form

Subject or Department _____

#	Date and Person Reporting	Specific Problem	Contributing Factors	Suggested Solution(s)	Responsible Department	Implementation Date	Results of Action	Next Review Date	Status and Further Action	Next Review Date	Status and Further Action	Date Solved

Basic Workers' Compensation Principles

Workers' compensation is a state-based system. Statute limits the compensation the worker receives, and the worker cannot sue in court for further damages. Key elements that are common to almost all states are as follows[5]:

1. All work-related injuries and illnesses must be compensated, regardless of fault.
2. Types of benefits:
   - *Temporary total.* Lost wages during recovery from a temporary injury or illness.
   - *Permanent total.* Lost wages for serious injury or illness where a return to work is not possible.
   - *Temporary partial.* Lost wage differential when a worker returns to work at lower wages.
   - *Permanent partial.* Future lost wage potential based on the residual of an injury or illness.
   - *Death.* Lost wages for a fatal injury paid to a surviving spouse and/or dependent children.
3. Medical benefits vary state to state. They are generally quite liberal, and the injured worker rarely faces out-of-pocket expenses.
4. Some of the major limitations of workers' compensation are that the system does not allow payments for pain and suffering, nor may workers receive punitive damages from their employers.

Workers' compensation is required in most states. Even in states where it is not required, most employers provide it. (An employer that does not carry coverage runs the risk of being sued for major legal damages.)

Types of Plans

Depending upon the state, there are three types of plans available, and employers must provide one of them:

1. *Commercial policies.* A medical practice may acquire a commercial policy (ie, from a for-profit insurance company) if the practice meets the state-determined workers' compensation requirements.
2. *Employer self-insured plans.* Employers may administer their own plans by setting aside funds, provided the coverage meets state requirements.
3. *State compensation fund.* Workers' compensation premiums are paid to the state agency that administers the plan.

Workers' Compensation Costs

The national annual average cost for an OSHA-recordable accident is approximately $25,000 to $30,000. An injury resulting from repetitive motion can run up to $100,000. Total workers' compensation costs continue to rise as the average cost of a claim soars. The costs of workers' compensation have risen steadily through the years, boosting the rates for both the medical and disability components of workers' compensation far greater than inflation.[6]

Figure 16.3 provides a formula for workers' compensation insurance premiums. The experience modification (ie, the *experience-mod*) is a critical factor in the final cost of an employer's workers' compensation policy. Much like the experience rating system used by many states to develop auto insurance rates, a bad year can haunt an employer for years to come. Insurance premiums are based on three consecutive years' workers' compensation experience.

By proactively avoiding costly workers' compensation claims, money is available that would otherwise go toward compensation expenses. As accountants and business owners know, it is not what the medical practice earns, but what it saves that counts.

The major "cost drivers" behind high compensation rates are rising medical costs, increased litigation, claims fraud, higher benefits, and a growing number of claims of long-term occupational illness and subjective injuries, such as stress.

Fraud has easily become the most visible issue in workers' compensation, with media reports of "injured" workers painting houses and moving furniture, unscrupulous revolving-door comp clinics, and indistinct attorneys exacting huge settlements for bogus claims. Fraud might account for up to 30% of total claims paid each year, according to some industry estimates, but the extent of the problem is unknown.

Workers' compensation reform legislation is spreading among states, and premium-reduction strategies are working. These include joint employer/employee safety committees, managed care programs, aggressive anti-fraud initiatives, strict limits on claims litigation, and the use of medical fee schedules. Some states have enacted stiff penalties for

FIGURE 16.3

A Formula for Workers' Compensation Insurance Premiums

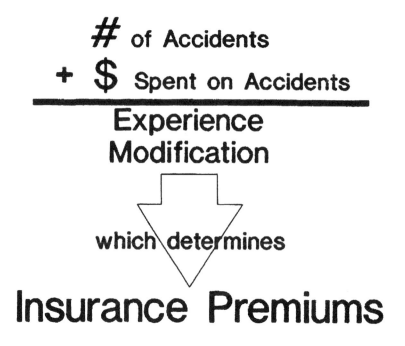

comp fraud, backed by media advertisements warning workers to "think twice" before filing false claims.

Uninsured Costs

Many experts believe the uninsured costs of on-the-job injuries and illnesses can be as much as two or three times the actual premium paid for workers' compensation insurance. Examples of costs not covered by workers' compensation insurance include the following:

- Replacement costs for temporary employees
- Production downtime
- Training temporary or replacement employees

Employee Incentives to Return to Work

Research shows that returning the employee to work as soon as medically possible is essential in properly managing workers' compensation claims, and it is better for the employee physically and psychologically. Benefits to employees include the following:

- Job site rehabilitation speeds recovery process.
- There is less disruption to employee's normal routine.
- Helps maintain social connections at work.
- Provides additional support during recuperation period.
- Employees realize their value to the medical practice.
- Employee tends to focus more on abilities instead of disability.
- Cultivates a stronger rapport among employees.
- Enhances the employer/employee relationship.

Management Incentives to Help Employees Return to Work

Generally speaking, an injured employee who cannot return to work is paid two thirds of his or her salary to recuperate at home. From a business standpoint, this expense means dollars that could have been spent on office equipment or employee benefits and raises. The following are benefits that supervisors and managers gain when they help an employee return to work as soon as possible:

- Saves money spent on hiring temporaries.
- Promotes positive employee morale.
- Utilizes recovery period to maintain productivity.
- Illustrates the value of the employee to the workplace.
- Promotes cross training.
- Minimizes potential for adversarial relationship among employees.
- Lowers medical costs.
- Decreases costs for lost workdays.
- Shortens case-management process.

- Minimizes short- and long-term disability cases.
- Promotes compliance with the Americans with Disabilities Act (ADA), when appropriate.

Workers' compensation represents a significant business expense for employers—in some cases, 20% to 30% of payroll. However, the system is critical to the well-being of the injured worker. There is a clear trend of increased benefits to the injured and inflated cost to the employer, which are passed on to consumers in terms of higher-priced medical care. One thing is certain—workers' compensation will cost more, not less, in the future.

To a large extent, employers can positively or negatively influence their net costs for workers' compensation coverage through their own efforts. This power is critical to the health of a medical practice, both in economic and human terms. There are sufficient cost incentives in the system to positively motivate management in that direction. The enormity of the cost and the means of controlling these expenses must be understood, accepted, and applied. When health and safety is a high priority, management, their employees, and the public realize the direct benefits.

There are also other potential advantages to medical practices having a good safety and health record. In states with assigned risk pools, rates are normally higher in the assigned risk pool than for medical practices that can obtain insurance in the voluntary market. This provides smaller medical practices that are not in an assigned risk pool an incentive to minimize illnesses and injuries so they are not dropped by their insurer and shuffled into the assigned risk pool. However, in some states many small medical offices, and virtually all new practices, are placed in the assigned risk pool in any case until they can provide a lower risk track record to insurance companies.

Some insurers may quote premiums below the prescribed rates for small medical practices. Such bargain policies tend to go to practices with good experience records. Premiums and benefits vary from one company to another, so carefully read the entire policy. It is quite possible that one company may offer lower premiums, but actually cost a medical practice more in the long run by not offering the same extent in coverage. It can be beneficial to ask the prospective insurer what can be done to possibly lower the premiums. Many insurers analyze the procedures or layout of a practice to determine if additional safety measures can be implemented that lower premiums.

Risk Management Strategies

Effective workers' compensation management requires a risk management approach to minimize the risks and reduce costs associated with employee injuries and illnesses. Risk management, as it relates to workers' compensation, has a two-part focus: prevention and case management.

Prevention
Prevention should be the main thrust of the strategies as described below:

- Develop and implement a safety program that fulfills the medical practices' moral, ethical, and legal responsibilities to its employees.
- Teach and encourage employees to participate in safety programs and processes.
- Focus on an active interview process to find an appropriate job fit when hiring.
- Train employees in their specific job responsibilities, including safety responsibilities.
- Establish an active, positive safety committee that involves supervisors and nonsupervisors.
- Educate employees about the workers' compensation system, including what it covers, what it does not cover, the cost of workers' compensation, the medical practice's philosophy about workers' compensation, and the return-to-work policies.

Case management
Case management is effective follow-up after employee incidents. Coordinate all of the factors and human elements to minimize human suffering and economic loss after the injury or illness as explained in the following sections.

Preplan Preventive strategies should minimize extensive case management activities. However, it is important to know how to effectively follow up when employee lost-time incidents occur. Employers have a responsibility to injured workers and to simultaneously control workers' compensation dollars. Case management steps include the following:

- *Appoint a coordinator.* Employers have greater power and control if they are prepared to deal with employee injuries and illnesses. Someone to act as a liaison between the employer, the injured employee, and outside agencies, such as the health care provider and insurance company, should be in charge of the Case Management Program.
- *Initiate a return-to-work program.* Transitional (eg, modified, light duty) jobs promote returning an employee to work as soon as possible. Table 16.2 outlines steps involved in a Return-to-Work Program.

  When employees return to work with restrictions and work in a transitional job, regular communication with the employees and their physicians shows concern and the desire to return them back to their full work schedules as soon as possible. An employee may be given permanent restrictions from performing regular duties. Rehabilitation consultants and claim representatives can become involved to facilitate placement efforts and other activities. The rehabilitation consultant can provide suggestions for keeping costs down while working through the work-restriction process.

  Another important step in case management is formalizing a relationship with a designated medical provider for occupational health services. Designated medical providers work with the employer and the employee with return-to-work efforts, are accessible for consultation, and should become familiar with the work environment and employer philosophies.

TABLE 16.2

Example of Return-to-Work Program Outline

Injuries requiring medical attention:

- Encourage employee to see a designated medical provider at least once.
- Contact the designated medical provider (or employee's physician) to obtain outcome of visit.
- Stress to physician willingness to offer transitional (eg, modified) job if physician places employee on restricted work duties.
- If physician does not release employee to return to work, maintain routine contact with employee at home. Let them know they are missed, send get-well cards, and so forth.
- Be aware of Plan of Care, and document discussions with employee and physician.
- Contact physician after each employee visit to discuss return-to-work projections and to discuss modified jobs available.

Physician releases employee to return to work with restrictions:

- Confirm restrictions in writing.
- Review existing pre-planned modified jobs for a match with the restrictions. If pre-planning was not done, review existing job duties. Some parts of the job may be able to be performed by employee.
- If necessary, combine available modified jobs with tasks in other departments. The job must be within the employee's restrictions.
- The ideal modified job has both positive and negative aspects just as the pre-injury job. Avoid placing unnecessary burden on employee or assigning harsh duties for the total modified job. Also avoid assigning employee the most desirable position in the office.
- Stick to previous shifts and schedules if possible.
- Send written suggested modified job to physician for approval.
- Once physician approves, present the transitional job to the employee. If mailing, send certified mail. Stress that the employee must not perform tasks outside of the modified job.
- If employee refuses the modified job, contact the claim representative. Negotiate changes to come to agreement.
- Maintain close contact with employee while on modified job. As employee's condition improves, be prepared to change the modified job until employee returns to pre-injury position.

Action Plan

Follow up on the injury or illness as soon as possible. Follow-up helps create positive workplace morale through communication and support for the injured worker and may reduce the potential for reinjury. The follow-up process also includes sending data to the medical practice's claims administration office that documents the incident and provides details for the investigation.

The employer should have an ongoing interaction with the injured employee to make sure that the employee understands the employer's concern, that treatment appointments are monitored, and that communications lines are kept open. This contact is important whether the employee is in the work environment in a transitional job or off work.

If there is a workers' compensation claim, an employer has the legal right to medical records specific to the work-related injury or illness. Establish lines of communication as soon as possible with the employee's medical provider. Contact with the employee's medical provider should convey concern regarding the employee's welfare, communicate the employer's return-to-work philosophy and expectations of cooperation with that philosophy, and emphasize the employer's willingness to work with the medical provider and the employee. The employer should also request a Plan of Care for the employee.

Signs of Potential Problems

Employers need to monitor the health care an employee receives after a work-related injury or illness. In cases of concern, contact the insurance company's claims representative to help develop a coordinated strategy to deal with the problems. Signs of potential problems include the following:

- No Plan of Care.
- Plan of Care not being followed.
- No improvement with treatment.
- Long-term use of pain medications or addictive medications.
- Lack of cooperation with return-to-work efforts.

Interface with Claims Representative

Many insurance companies provide claims representatives to assist employers with workers' compensation claims. These individuals are valuable resources and should be utilized whenever possible. Any time the employer is aware of a change in the employee's status, whether related to the claim or employment, notify the claims representative.

Complying with Americans with Disabilities Act

Employers often ask about responsibilities under the Americans with Disabilities Act (ADA) to workers injured on the job. Most employees injured on the job will not meet the ADA's definition of "disabled." Only injured workers who meet the ADA's definition of an "individual with a disability" and "qualified individual with a disability" are considered disabled under the ADA, regardless of whether they receive workers' compensation benefits. Work-related injuries and illnesses in medical practices do not usually cause physical or mental impairments severe enough to "substantially limit" a major life activity or are of such short duration, with little or no long-term effect (ie, broken arm). Such injuries and illnesses, in most circumstances, are not considered disabilities under the ADA.

An employee who injures her back on the job and is prohibited only from lifting extremely heavy objects probably does not have a "disability" under the ADA even though she may have a vocational disability for purposes of the Workers' Compensation Law. However, if the injury results in a condition that qualifies as a "disability," employers are required to make "reasonable accommodation," which may include leave. Where an injured employee meets the ADA's definition of "disabled," the employer must attempt to reasonably accommodate the employee using the statutory guidance of the ADA.

REPORTING REQUIREMENTS

Incident reporting requirements can be met by following the OSHA Recordkeeping Requirements and the Safe Medical Device Reporting Act (SMDR) of 1990. Some medical practices provide reports of occupational injuries and illnesses to their insurance providers, and the OSHA reports can help simplify this process.

OSHA Recordkeeping Guidelines

OSHA's Recordkeeping Guidelines (Standard Number 29 CFR 1910.1904) requires covered employers to prepare and maintain records of occupational injuries and illnesses. These requirements provide the data for the BLS Occupational Injury and Illness Survey, the primary source of statistical information concerning workplace injuries and illnesses. BLS collects the data and publishes the statistics, while OSHA interprets and enforces the regulation. These requirements include forms to complete, information on where to keep records, how to determine when to record an injury or illness, and a summary of checks and balances built into the system.

In 2001, OSHA issued a revised rule to improve the system employers use to track and record workplace injuries and illnesses. The new rule is expected to be passed into law. The revision improves employee involvement, creates simpler forms, provides clearer regulatory requirements, and allows employers more flexibility for using computers to meet OSHA regulatory requirements. Some industries classified as low-hazard sectors are exempt from most of the revised rule. When passed, medical practices should consult the revised rule for compliance requirements.

This section does not cover OSHA requirements for maintenance and retention of records for medical surveillance, exposure monitoring, inspections, and other activities and incidents relevant to occupational safety and health, and for the reporting of certain information to employees and to OSHA. For information on those requirements, refer directly to OSHA (Standard Number 29 CRF 1910.1020 *Access to Employee Exposure and Medical Records*).[7] The following sections offer an overview of the current OSHA recordkeeping guidelines.

Concepts of Recording

Instructions on the back of OSHA Form 200 include basic recordkeeping concepts and guidelines. Recordable and nonrecordable injuries and illnesses are distinguished by the treatment provided (ie, if the injury is such that medical treatment was provided or should have been provided, it is recordable; if only first aid was required, it is not recordable). However, medical treatment is only one of several criteria for determining recordability. Regardless of treatment, if the injury involved loss of consciousness, restriction of work or motion, or transfer to another job, the injury is recordable. Table 16.3 lists procedures generally considered medical treatment if the injury is work related.

First-aid treatment is one-time treatment and subsequent observation of minor injuries and should not be recorded if the work-related injury does not involve loss of consciousness, restriction of work or motion, or transfer to another job. Table 16.4 lists procedures generally considered first-aid treatment.

Some medical practices and other health care facilities also use alternate surveillance systems to record percutaneous injuries involving exposure to bloodborne pathogens because the OSHA 200 log does not accurately reflect needlestick injuries. The OSHA 200 log does not require recordkeeping for all needlestick injuries that pose a potential risk of disease transmission, such as percutaneous injuries that affect or pass through the skin. Alternate recording systems vary in content and level

TABLE 16.3

A List of Generally Considered Medical Treatments

- Treatment of INFECTION
- Application of ANTISEPTICS during second or subsequent visit to medical personnel
- Treatment of SECOND- OR THIRD-DEGREE BURN(S)
- Application of SUTURES (stitches)
- Application of BUTTERFLY ADHESIVE DRESSING (S) or STERI STRIP (S) in lieu of sutures
- Removal of FOREIGN BODIES EMBEDDED IN EYE
- Removal of FOREIGN BODIES FROM WOUND; if procedure is COMPLICATED because of depth of embedment, size, or location
- Use of PRESCRIPTION MEDICATIONS (except a single dose administered on first visit for minor injury or discomfort)
- Use of hot or cold SOAKING THERAPY during second or subsequent visit to medical personnel
- Application of hot or cold COMPRESS (ES) during second or subsequent visit to medical personnel
- CUTTING AWAY DEAD SKIN (surgical debridement)
- Application of HEAT THERAPY during second or subsequent visit to medical personnel
- Use of WHIRLPOOL BATH THERAPY during second or subsequent visit to medical personnel
- POSITIVE X-RAY DIAGNOSIS (eg, fractures, broken bones)
- ADMISSION TO A HOSPITAL or equivalent medical facility FOR TREATMENT

Reproduced from OSHA Record Keeping Summary, OSHA Record Keeping Guidelines (29 CFR Part 1904).

TABLE 16.4

A List of Procedures Generally Considered First-Aid Treatments

- Application of ANTISEPTICS during first visit to medical personnel
- Treatment of FIRST-DEGREE BURN (S)
- Application of BANDAGE (S) during a visit to medical personnel
- Use of ELASTIC BANDAGE (S) during first visit to medical personnel
- Removal of FOREIGN BODIES NOT EMBEDDED IN EYE if only irrigation is required
- Removal of FOREIGN BODIES FROM WOUND; if procedure is UNCOMPLICATED, and is, for example, by tweezers or other simple technique
- Use of NONPRESCRIPTION MEDICATIONS AND administration of single dose of PRESCRIPTION MEDICATION on first visit for minor injury or discomfort
- SOAKING THERAPY on initial visit to medical personnel or removal of bandages by SOAKING
- Application of hot or cold COMPRESS (ES) during first visit to medical personnel
- Application of OINTMENTS to abrasions to prevent drying or cracking
- Application of HEAT THERAPY during first visit to medical personnel
- Use of WHIRLPOOL BATH THERAPY during first visit to medical personnel
- NEGATIVE X-RAY DIAGNOSIS
- OBSERVATION of injury during visit to medical personnel

Administration of TETANUS SHOT (S) or BOOSTER (S), by itself, is not considered medical treatment. However, these shots are often given in conjunction with more serious injuries; consequently, injuries requiring these shots may be recordable for other reasons.

Reproduced from OSHA Record Keeping Summary, OSHA Record Keeping Guidelines (29 CFR Part 1904).

of details and include reports submitted through the Safe Medical Device Act.

The Mechanics of OSHA Recordkeeping

The two forms used for the current OSHA recordkeeping requirements are: OSHA 200 and OSHA 101.

Log of occupational injuries and illnesses, OSHA 200 The following summarizes the major concepts of OSHA 200 log reporting requirements:

- An injury or illness is considered work related if it results from an event or exposure in the work environment. The work environment is primarily composed of the employer's premises and other locations where employees are engaged in work-related activities or are present as a condition of their employment.
- All work-related fatalities are recordable.
- All recognized or diagnosed work-related illnesses are recordable.

The OSHA 200 log serves two purposes:

1. As the Log of Occupational Injuries and Illnesses in which the occurrence, extent, and outcome of cases are recorded during the year.
2. As the Summary of Occupational Injuries and Illnesses used to summarize the log at the end of the year to satisfy employer posting obligations (must be posted annually from February 1 to March 1).

The OSHA 200 log records and classifies recordable occupational injuries and illnesses and notes the extent and outcome of each case. The log shows the following:

- When the occupational injury or illness occurred.
- To whom.
- What the injured or ill person's regular job was at the time of the injury or illness exposure.
- The department in which the person was employed.
- Type of injury or illness.
- How much time was lost.
- Whether the case resulted in a fatality.

Part 1904.2[8] of these guidelines provides the requirements for the log. The log consists of three parts:

- A descriptive section that identifies the employee and briefly describes the injury or illness.
- A section covering the extent of the injuries recorded.
- A section on the type and extent of illnesses.

Most of the columns are self-explanatory. *OSHA Record Keeping Guidelines* provide information to assist in completing the log.

Supplementary record of occupational injuries and illnesses, OSHA 101

The OSHA 101 is the Supplementary Record of Occupational Injuries and Illnesses and provides additional information on each of the cases that have been recorded on the log. Figure 16.4 is a sample of the OSHA 101 form.

FIGURE 16.4

The Occupational Safety and Health Administration Supplementary Record of Occupational Injuries and Illnesses
Reproduced from OSHA Record Keeping Guidelines (29 CFR Part 1904).

Occupational Safety and Health Administration
Supplementary Record of
Occupational Injuries and Illnesses

U.S. Department of Labor

This form is required by Public Law 91-596 and must be kept in the establishment for 5 years. Failure to maintain can result in the issuance of citations and assessment of penalties.

Case or File No.

Form Approved
O.M.B. No. 1218-0176
See OMB Disclosure Statement on reverse.

Employer

1. Name
2. Mail address (No. and street, city or town, State, and zip code)
3. Location, if different from mail address

Injured or Ill Employee

4. Name (First, middle, and last)
   Social Security No.
5. Home address (No. and street, city or town, State, and zip code)
6. Age
7. Sex (Check one) Male ☐ Female ☐
8. Occupation (Enter regular job title, not the specific activity he was performing at the time of injury.)
9. Department (Enter name of department or division in which the injured person is regularly employed, even though he may have been temporarily working in another department at the time of injury.)

The Accident or Exposure to Occupational Illness

If accident or exposure occurred on employer's premises, give address of plant or establishment in which it occurred. Do not indicated department or division within the plant or establishment. If accident occurred outside employer's premises at an identifiable address, give that address. If it occurred on a public highway or at any other place which cannot be identified by number and street, please provide place references locating the place of injury as accurately as possible.

10. Place of accident or exposure (No. and street, city or town, State, and zip code)
11. Was place of accident or exposure on employer's premises? Yes ☐ No ☐
12. What was the employee doing when injured? (Be specific. If he was using tools or equipment or handling material, name them and tell what he was doing with them.)
13. How did the accident occur? (Describe fully the events which resulted in the injury or occupational illness. Tell what happened and how it happened. Name any objects or substances involved and tell how they were involved. Give full details on all factors which led or contributed to the accident. Use separate sheet for additional space.)

Occupational Injury or Occupational Illness

14. Describe the injury or illness in detail and indicate the part of body affected. (E.g., amputation of right index finger at second joint; fracture of ribs; lead poisoning; dermatitis of left hand, etc.)
15. Name the object or substance which directly injured the employee. (For example, the machine or thing he struck against or which struck him; the vapor or poison he inhaled or swallowed; the chemical or radiation which irriatated his skin; or in cases of strains, hernias, etc., the thing he was lifting, pulling, etc.)
16. Date of injury or initial diagnosis of occupational illness
17. Did employee die? (Check one) Yes ☐ No ☐

Other

18. Name and address of physician
19. If hospitalized, name and address of hospital

Date of report | Prepared by | Official position

(See Next Page/Reverse)

FIGURE 16.4

continued

**SUPPLEMENTARY RECORD OF OCCUPATIONAL
INJURIES AND ILLNESSES**

To supplement the Log and Summary of Occupational Injuries and Illnesses (OSHA No. 200), each establishment must maintain a record of each recordable occupational injury or illness. Worker's compensation, insurance, or other reports are acceptable as records if they contain all facts listed below or are supplemented to do so. If no suitable report is made for other purposes, this form (OSHA No. 101) may be used or the necessary facts can be listed on a separate plain sheet of paper. These records must also be available in the establishment without delay and at reasonable times for examination by representatives of the Department of Labor and the Department of Health and Human Services, and States accorded jurisdiction under the Act. The records must be maintained for a period of not less than five years following the end of the calendar year to which they relate.

Such records must contain at least the following facts:

1) About the employer - name, mail address, and location if different from mail address.

2) About the injured or ill employee - name, social security number, home address, age, sex, occupation, and department.

3) About the accident or exposure to occupational illness - place of accident or exposure, whether it was on employer's premises, what the employee was doing when injured, and how the accident occurred.

4) About the occupational injury or illness - description of the injury or illness, including part of the body affected, name of the object or substance which directly injured the employee; and date of injury or diagnosis of illness.

5) Other - name and address of physician; if hospitalized, name and address of hospital, date of report; and name and position of person preparing the report.

SEE *DEFINITIONS* ON THE BACK OF OSHA FORM 200.

OMB DISCLOSURE STATMENT

Public reporting burden for this collection of information is estimated to average 20 minutes per response, including the time for reviewing instructions, searching existing data sources, gathering and maintaining the data needed, and completing and reviewing the collection of information. Persons are not required to respond to the collection of information unless it displays a currently valid OMB control number. If you have any comments regarding this estimate or any other aspect of this information collection, including suggestions for reducing this burden, please send them to the OSHA Office of Statistics, Room N3644, 200 Constitution Avenue, NW, Washington, DC 20210

DO NOT SEND THE COMPLETED FORM TO THE OFFICE SHOWN ABOVE

For every injury or illness entered on the log, it is necessary to record additional information on the Supplementary Record, OSHA 101. The Supplementary Record describes how the injury or illness exposure occurred, lists the objects or substances involved, and indicates the nature of the injury or illness and the part(s) of the body affected. Part 1904.4 of the guidelines[8] provides the requirements for the Supplementary Record and is summarized as follows:

- Supplementary Records must be available for inspection within 6 working days after receiving information that a recordable case has occurred.
- The instructions must be followed when completing the record.
- Workers' compensation insurance or other reports are acceptable alternative records if they contain the information required by the OSHA 101 form.

Location, Retention, and Maintenance of Records
OSHA regulations require that records be located and maintained for the medical practice for several reasons. Records assist government agencies in administering and enforcing the act, increase employer–employee awareness, and promote injury and illness prevention.

Access to OSHA Records
The integrity of the recordkeeping process must be ensured, including access to records. OSHA *Record Keeping Guidelines* describe requirements concerning access to OSHA injury and illness records, including the following:

- Which OSHA records are subject to the access provisions.
- What is meant by the term "access."
- Whether employees can gain access to any other injury and illness records.
- If employees can see the entire log or only that portion containing an entry that specifically relates to them.

Penalties for Failure to Comply with Recordkeeping Obligations
OSHA prescribes penalties for the falsification of OSHA records, the failure to keep the OSHA records, or the failure to make OSHA reports. Employers violating recordkeeping requirements are subjected to the same sanctions as other safety and health standards and regulations. Consequently, the agency vigorously pursues recordkeeping and reporting violations to ensure the continued integrity of the records and validity of the data that is produced.

Safe Medical Device Act of 1990

Medical device reporting (MDR) and manufacturers and user facility device experience (MAUDE) are a means for the FDA, manufacturers, and user facilities to identify and monitor significant adverse events that involve medical devices. These reports are used to efficiently detect and correct problems.

Under the SMDA, a device is classified as anything that is used in the diagnosis or treatment of a patient *other* than drugs. However, because these devices could just as easily be hazardous to employees, their potential for harm must be viewed from both patient and employee perspectives.

ENDNOTES

1. National Safety Council. *Injury Facts 2000*. National Safety Council: Itasca, IL; 2000.
2. DOL *News*. US Department of Labor, Bureau of Labor Statistics: Washington, DC. March 28, 2001.
3. National Institute for Occupational Safety and Health. *Worker Health Chartbook 2000*. US Department of Health and Human Services, National Institute for Occupational Safety and Health: Cincinnati, OH. DHHS (NIOSH) Publication Number 2000-127.
4. National Institute for Occupational Safety and Health. *Worker Health Chartbook 2000*. US Department of Health and Human Services, National Institute for Occupational Safety and Health: Cincinnati, OH. DHHS (NIOSH) Publication Number 2000-127.
5. Gice J. *The Cost of Comp*. Occupational Health & Safety. Dallas, TX: Stevens Publishing Corporation; 2001:59-60.
6. Menard RA., II, *Talking Dollars & Sense*. Occupational Health & Safety. Dallas, TX: Stevens Publishing Corporation; 2001:62-65.
7. Occupational Safety and Health Administration Regulation (Standard Number 29 CRF 1910.1020, *Access to Employee Exposure and Medical Records*).
8. Occupational Safety and Health Administration Regulation (Standard Number 29 CFR 1910.1904, *Recordkeeping Regulations*).

chapter 17

Safety Program Evaluation

Setting up a safety program is the first step. Evaluating and maintaining the program then become ongoing steps in the process.

Once the safety and health program is in place, it must be constantly evaluated to ensure that it is achieving its established goals. This process of evaluation not only increases the program's effectiveness, but also helps to easily and effectively implement the new, ongoing changes in regulations.

The following sections describe how to develop an ongoing evaluation process for the medical practice's safety program. It also describes how to use the safety program evaluation as a tool for resource allocation.

ONGOING PROGRAM EVALUATION

Developing an evaluation process involves the following six steps:

1. Identify program elements
2. Select specific indicators
3. Develop a sampling plan
4. Collect the data
5. Analyze the data
6. Implement corrective actions

Identify Program Elements

The first step in developing the evaluation process is to identify all the safety and health elements that need to be monitored, such as:

- Program administration
- Information management
- Infection control
- Hazardous chemicals
- Safety promotion and motivation
- Medical equipment
- PPE
- Hazardous and medical waste management
- Personal safety and health

- Ergonomics
- Workplace violence
- Indoor air quality
- Patient safety
- Emergency management and response
- Education and training
- The effect of accidents and illnesses

Select Specific Indicators

The second step in developing the evaluation process is to select specific indicators that measure each element's effectiveness and to signal when corrective action needs to be taken. Usually, three indicators provide a basis for evaluation; however, additional indicators may be needed depending on the type of program and its success. The following are the three key elements of an evaluation:

1. *Policies and procedures.*
   - Are policies and procedures in place?
   - Are they current?
   - Do observations show that employees are following the procedures?
2. *Training.*
   - Do records show that all employees are trained?
   - Do employees demonstrate that they know the procedures through either written or verbal testing?
   - Is a process in place to retrain employees who do not know the procedures?
3. *Ongoing monitoring and corrective action.*
   - Is a process in place to monitor the performance of all programs?
   - Have indicators and monitoring frequencies been established based on sound statistical procedures?
   - Is there documentation and a tracking system of corrective action elements, responsibility, and timing for identified safety problems?

Develop a Sampling Plan

The third step in developing the evaluation process is to devise a sampling plan that specifies what information should be collected, how much information to collect, and how to collect it. Because this process requires data management skills, it is important that the individuals responsible for establishing and running the program be trained in the statistical techniques of data collection. In addition, they should be trained in the use of computerized data management and analysis. Sound statistical data analysis allows the practice to respond to safety problems and to take corrective action to bring about change in safety performance, thereby avoiding reacting to data that is not statistically significant. Further, using a statistical approach identifies those elements of the safety program that can be measured and those that cannot. Many indicators provide insufficient data to allow statistical analysis during a short time span. In these cases, common sense and experience may be more effective in determining if corrective action is necessary. Even when the results are not statistically significant, they should be examined for obvious trends.

Collect the Data

The fourth step in the evaluation process is to actually collect the data, via employee interviews and testing, which is based on information developed in the first three steps. Unannounced walk-through inspections by safety personnel and safety committee members can also be effective in uncovering equipment hazards and unsafe practices. Employees should be asked about their reactions to the practicality and effectiveness of various program elements. Additionally, department managers and supervisors can monitor daily practices. When employees are not working safely, managers should determine why, assuring employees that they want to know if the problem is one of poorly designed procedures or insufficient training. Managers and supervisors can explain that employee input is valuable to the sustained success of the safety program. This communication is essential because line employees are most responsible for daily implementation of the program. Another way to assess employee reactions to procedures is through the use of employee response to questions on in-service evaluation forms. Finally, in addition to discussions with employees, program effectiveness can be evaluated by observing trends in incidents, injuries, and illnesses. To the extent possible, every effort should be made to collect objective data.

Analyze the Data

The fifth step in program evaluation is to analyze the data in order to identify specific problems and trends that require corrective action. The results of the analysis are then furnished to those who are responsible for taking corrective action. In addition, the information is summarized and given to the safety committee for its use in supervising and managing the overall safety program. Summaries of the indicators and corrective actions taken are compiled and issued by the safety committee in minutes and quarterly reports.

Implement Corrective Action

Once the data has been analyzed, it can then be reviewed to decide what corrective action needs to be taken. When employees realize that the results of their input affect corrective actions, employee confidence and participation in the process increase.

ANNUAL PROGRAM EVALUATION

In addition to ongoing program evaluation, the practice must evaluate the overall safety program on an annual basis. The safety committee usually develops this evaluation as a report. The report shows progress and accomplishments using monitoring data and information from all the committee and subcommittee reports. It should spell out what the previous goals were, and if they were met. It should also specify the goals for the coming year.

The following are a number of sources of information that may be used to prepare the report:

- Incident and illness reports
- Lost-time incident reports
- Voluntary and regulatory agency survey reports
- Insurance company inspection reports
- Training records
- Workers' compensation statistics
- New and old policies and procedures

The annual report covers all elements of the safety and health program. The report is vital because it provides a well-documented analysis of the effectiveness of the total safety management program and specifies the medical practice's performance improvement strategy.

EVALUATION OF NEW AND REVISED SAFETY PROGRAMS

Although ongoing evaluation is usually adequate for safety programs that are proven to be effective, a more complete evaluation should be undertaken 6 months after implementation of a new or revised program. The purpose of this evaluation is to ensure that new procedures have been fully implemented, that they are accomplishing their intent, and that they are achieving their goals in an acceptable manner. If the program has been revised as a result of a needs assessment, the review should utilize the results of the original needs assessment as a basis for determining progress.

Even though the first evaluation should take place approximately 6 months after implementation of a new program element, dramatic results should not be immediately expected. Developing and implementing the policies and procedures, introducing them to staff, and training employees in carrying them out may require reevaluation and development of new performance goals.

Without the tools of evaluation and consistent monitoring, any program could suffer after the initial push for implementation. Ongoing review and measurement of results is essential in determining what the program has accomplished and where future efforts should be focused.

PROGRAM EVALUATION AS A TOOL FOR RESOURCE ALLOCATION

Most facilities face an ongoing problem of balancing safety management against other demands. For example, every practice is faced with internal and external demands on its resources, such as:

- Cost reduction
- Rapid growth
- Increased government regulation

Frequently, safety resources, both staffing and finances, are cut because of the lack of measurable accomplishments. In addition, the consequences of meeting safety regulatory requirements are not often adequately communicated to senior management.

With growth, there may be major changes in services or other processes that make some policies and procedures obsolete and may necessitate major revisions in existing procedures. For example, as the demographics of society change, medical practices may move toward expanded services for the elderly whose needs differ from those of younger patients. In addition, growth means the use of equipment that relies on advanced technology, requires stepped-up training for users, needs stricter safety monitoring, and necessitates preventive maintenance.

Government regulations constantly change and, thus, require alterations in practices that once were routine. Stricter enforcement and penalties make adherence to new standards even more important because fines will affect the practice's earnings. Many regulations are complex and require significant resources to implement and maintain compliance.

Because of these competing resource demands, safety program management must keep pace by using the best safety management technology and processes available. An important benefit of program evaluation is that it can define the most effective use of available resources. The knowledge gained through evaluation enables the safety committee to prioritize resource allocation to programs that are not meeting goals. In addition, where there are insufficient resources allocated to correct existing program deficiencies and to implement new programs, information from the evaluation process provides the data to support the need for additional resources. One of the responsibilities of the safety committee is to ensure that those involved in resource allocation are aware of the consequences of cutting back on specific program support.

BENCHMARKING SAFETY AND HEALTH PROGRAMS

An emerging practice in the arena of occupational health and safety is known as *benchmarking*. The processes described throughout this book parallel those described by advocates of benchmarking projects. Whether efforts to enhance employee safety and health are called a safety program, a health and safety program, or a health and safety benchmarking project is irrelevant, the outcome of the efforts will be the same. The medical practice will realize more efficient and effective operations and a safer, healthier, and happier workforce. This is the ultimate goal.

appendix

About the CD-ROM

The CD-ROM in the back of this book contains the full text of the standards, recommendations, guidelines, and other material referenced in this book, organized by chapter. By clicking a chapter folder on the CD-ROM, you will open a list of the documents used to create the information in that particular chapter. Questions about the content of each document should be directed to the agency or organization that produced the material. To install the CD-ROM, follow these steps:

1. Insert the CD-ROM into your CD-ROM drive.
2. Double-click the CD-ROM icon to open the directory. (If the CD-ROM icon is not on your desktop, select Start, Run, and type in the letter of your CD-ROM drive, and click OK.)

All files are Adobe Acrobat files. To open, you must have the free Adobe Acrobat Reader software installed on your computer. If you do not have Adobe Acrobat Reader installed, you may download it at the Adobe website at www.adobe.com.

index

A

Accidents. *See also* Incidents; Injuries; Workers' compensation
  eliminating, 285
  risk management strategies for, 290–292
Aerosolization, 34
Agencies. *See* Federal agencies; Government compliance agencies
Aging, ergonomic recommendations and, 270
Aging workers, in medical practices, 267–270. *See also* Multicultural employees; Young workers
Aisleways, patient safety in, 211
Alternative keyboards, 172–173
Alternative work arrangements, 277
American Conference of Governmental Industrial Hygienists, Inc. (ACGIH), 25
American National Standards Institute (ANSI), 28
American Society of Heating, Refrigerating, and Air-Conditioning Engineers (ASHRAE), 25–26
Americans with Disabilities Act (ADA), 21–22
  federal agencies responsible for, 21
  medical practices and, 21–22
  on-job injuries and, 293
Antineoplastic drugs, 62
Aprons, 118
Argon, safety guidelines for, 93
Arson, 226
Assault, 176. *See also* Workplace violence
Atmospheric gases, odorizing, 94
Attendance records, 10

B

Back pain injuries, 149
Benchmarking, 305
Bloodborne pathogens. *See also* Infections
  consequences of exposures to, 279
  dealing with exposure to, 50–51
  preventing exposure to, 37
  recording exposure to, 294–296
  reducing risk of exposure to, 49–50
  universal precautions for, 44
  work practice controls for, 47
Bloodborne Pathogens (BBP) Standard, 41–51
  assessing type of exposure, 45–46
  compliance methods for, 44–49
  engineering and work practice controls, 47
  Exposure Control Plan, 42–44
  glove requirements, 117–118
  guidelines for communicating hazards to employees, 49
  hand-washing facilities and practices and, 48
  HBV, postexposure program, and follow-up, 48
  housekeeping standards for, 48
  laundry and, 48
  personal protective equipment and, 47–48
  recordkeeping and, 49
  requirements of, 41–43
Body movement, basic principles of, 150
Body protection, 118
Body Substance Isolation (BSI), 51
Bomb threats, 179, 239–240. *See also* Emergency management
  response form for, 241
Brainstorming, as training technique, 259–260
Building-related illnesses, 195. *See also* Indoor air quality (IAQ)
  HVAC systems and, 196–197
Bureau of Labor Statistics (BLS), 278

C

Capital equipment. *See* Equipment
Carbon dioxide (carbonic acid gas), safety guidelines for, 90–91
Carbonic acid gas (carbon dioxide), safety guidelines for, 90–91
Carpel tunnel syndrome, preventing, 166
Case studies, as training technique, 260
Ceiling, 60
Center for Devices and Radiological Health (CDRH), 18–19
Centers for Disease Control and Prevention (CDC), 22–23
Chain of command, for emergencies, 242
Change, safety and health programs, 151
Chemical Hygiene Plan, 73
Chemicals
  abbreviations for exposure values for, 60–61
  employee reactions to, 71–72
  entrance routes into human body for, 59–60
  hazardous, 16
  hazardous, in laboratories, 72–73
  in medical practice, 61–62
  routes of entry to body, 59–61
  working with, 70–71
Chemotherapeutic drugs, 62–64
Children, patient safety and, 215–217
Clean Air Act (CAA), 16
Clean Water Act (CWA), 16
Clinical practices. *See* Medical practices
Codman, Ernest, 26
Commitment, safety programs and, 3, 10–11
Communication, during emergencies, 242
Comprehensive Environmental Response, Compensation, and Liability Act (CERCLA or Superfund), 16
Compressed Gas Association, 28–29
Computer-based training, 262–263
Containers. *See* Cylinders; Disposal containers
Contaminated laundry, BBS Standard and, 48
Contaminated material, contact with, 35
Controlled Substances Act (CSA), 19
Corporate "virtual" universities, 263
Council of State Governments (CSG)
  medical waste guidelines
  containment, 137–138
  contingency planning, 141–143
  handling, 137
  labeling, 139
  monitoring and recordkeeping, 141
  segregation, 137
  storage prior to treatment, 140–141
  training, 141
  transportation, 143
Cumulative trauma disorders (CTDs), 160
Current employee level of training, 265
Customer outreach programs, 14–15
Cylinders, medical gas, 82–83
  handling and storing guidelines for, 83–84
  leak test procedures for, 84–87
  moving, 87
  opening and closing values, 88
  personal protective equipment for, 84
  receiving, 84
  safety standards for, 83
  storage requirements for, 85–86
  storing guidelines, 87–88
Cytostatic drugs, 62–64

309

Cytotoxic drugs, 62–64
Cytotoxic waste. *See also* Waste, healthcare
   disposing, 131
   reducing, 133–134

D

Data evaluation, 106–107
Data-tracking programs, 285, 286
Deaths, work-related, 277
Defense mechanisms, for infections, 35
Delusional stalkers, 182
Demonstrations, as training technique, 259
Department of Transportation (DOT), 19
Devices. *See* Equipment
Disasters. *See* Natural disasters
Discussion, as training technique, 259
Disposal containers
   labeling, 139
   recommended practices for, 138
   for sharps and needles, 40
Disposal contractors, choosing, 143–146
Distance learning, 263
Domestic violence, 183–184. *See also* Workplace violence
Drug Enforcement Administration, 19

E

Earthquakes, 235–236
Emergencies. *See* Bomb threats; Hazards; Natural disasters; Technological emergencies
Emergency management
   core operations of, 223–224
   patient safety and, 218
   planning process for, 222
   positive aspects of, 221–222
Emergency planning, OSHA regulations for, 240–245
Emergency response teams, 243
Employee attendance records, 10
Employee involvement, promoting and securing, 147–148. *See also* Safety and health committees
Employee perceptions, of safety programs, 6
Employees
   aging, 267–270
   multicultural, 273–275
   young, 269–273
Employee training
   for emergencies, 244
   for personal protective equipment (PPE), 119–120
   testing effectiveness of, 7
Engineering controls
   for bloodborne pathogens, 47
   for reducing workplace hazards, 156
Entrances
   patient safety in building, 209–210
   patient safety in office, 210
Environmental impairment liability, 127
Environmental Protection Agency (EPA), 16–17, 22
Environmental tobacco smoke (ETS), 202
EPA. *See* Environmental Protection Agency (EPA)
Equipment
   assessing condition of, 7
   JCAHO management programs for, 101
   OSHA standards for hazardous, 100
   procurement planning for, 102
   purchasing capital, 103
   servicing, 94–95
   types of, in medical practices, 101–102
Ergonomics
   applying, in medical offices, 162–170
   commonplace terms for, 160
   defined, 159–160
   developing programs for, 164–168
   identifying, with incident reports, 169–170
   identifying problems of, 169
   OSHA and, 174
   overt workplace injuries and, 161
   prevalency of, 161
   recommendations, and aging, 270
   resources for, 172–174
   terminology glossary for, 160 (table)
   in workplaces, 161–162
Examination rooms, patient safety in, 212
Exhaling, 150
Exits, patient safety in, 213
Exposure Control Plan, for bloodborne pathogens, 42–44
Exposure dose, definition of, 60
Exposure Prevention Information Network (EPINet), 37
Extinguishers, fire, 227–228, 229
Eye protection, 114–115

F

Face protection, 114–116
Facilitators, safety and health committee, 148–149
Facilities, health care
   assessing physical condition of, 5–6
   patient safety inside of, 210–213
   patient safety outside of, 208–210
Falls
   patient safety and, 214–215
   preventing, 96–97
FDA Desk Guide for Reporting Adverse Events and Product Problems (FDA), 18–19
Federal agencies. *See also* Government compliance agencies; Nongovernmental organizations
   Centers for Disease Control and Prevention, 22–23
   National Institute for Occupational Safety and Health, 23
Federal Insecticide, Fungicide, Rodenticide Act (FIFRA), 16
Fire extinguishers, 227–228, 229
Fire Protection Handbook (NFPA), 28
Fires, 224–226. *See also* Emergency management; Natural disasters
   being trapped in burning buildings, 231
   evacuation procedures for, 230–231
   extinguishers for, 227–228, 229
   preventing, 226–227
   responding to, 229–230
   surviving after, 231–233
   training and drills for, 228–229
First-aid treatments, 294–295
Floods, 233–234
Food and Drug Administration (FDA), 17–19
Foot protection, 119
Formaldehyde, 64–66
Formaldehyde waste, reducing, 134. *See also* Waste, healthcare
Formal investigations, 282–283

G

General Duty Clause, 15
Gloves, 116–118. *See also* Hand protection
   checklist for, 117
   latex allergies and, 120–122
Glutaraldehyde, 66–67
Government compliance agencies. *See also* Federal agencies; Nongovernmental organizations
   for Americans with Disabilities Act (ADA), 21
   Department of Transportation (DOT), 19
   Drug Enforcement Administration, 19
   Environmental Protection Agency (EPA), 16–17, 22
   Food and Drug Administration (FDA), 17–19
   Nuclear Regulatory Commission, 20–21
   Occupational Safety and Health Administration (OSHA), 13–16
   Office of Justice Programs, 19–20
   Principles of Good Regulation, 20
   state and local, 24
Group brainstorming, as training technique, 259–260

H

Half-life, 20
Hallways, patient safety in, 210
Hamilton, Alice, 13
Hand protection, 116–118. *See also* Gloves
Hazard assessment
  certification of PPE (sample), 113
  guidelines, 112
  protective equipment selection for, 111–114
Hazard Communication Standard (HAZCOM), 15
  complying with, 75–77
  requirements, 73–75
Hazard evaluation, 8
Hazardous chemicals. *See* Chemicals
Hazardous Materials Safety, Office of, 19
Hazards. *See also* Emergency management
  arson, 226
  control methods for reducing workplace, 154–157
  fires. *See* Fires
  setting priorities for, 9
Head protection, 118–119
Health and safety programs. *See* Safety and health programs
Healthcare waste. *See* Waste, healthcare
Health care workers. *See also* Medical practices
  infections in, 278–279
  risk of work-related assaults of, 175–176
  stalking of, 180–182
  threats to, 179–180
  workplace violence and, 177
Hearing protection, 119
Hospitals, establishing safety programs by, 1–2
Housekeeping, for BBS Standard, 48
Hurricanes, 238–239
HVAC systems, building-related illnesses and, 196–197

I

Illnesses, reporting, 279–281
Immediately Dangerous to Life and Health (IDLH), 60
Incident reports
  employee participation in, 280
  for ergonomic problems, 169–170
  filing, 279–280
  information to be included in, 280–281
  routing, 281
Incidents
  case management for, 291
  follow-ups for, 292–293
  improvement goals for, 285
  investigating, 281–283
  preventive strategies for, 290–291
  reporting requirements, 293–300
Indoor air quality (IAQ), 193–194
  basic precautions for improving, 203
  common complaints related to, 194–195
  external building activities and, 197–198
  housekeeping and maintenance procedures for improving, 200–201
  internal building activities and, 198–199
  multiuse buildings and, 203
  new construction and, 203–204
  pest control and, 201
  programs for, 204–205
  renovation or remodeling activities and, 202
  smoking and, 202
  waste management and, 201
Infections. *See also* Bloodborne pathogens; Needlestick injuries; Tuberculosis (TB)
  defense mechanisms for, 35
  in health care workers, 278–279
  program elements for controlling, 36–37
Infectious diseases, 1
  Body Substance Isolation (BSI) for, 51–52
  defense mechanisms for, 35
  examples of, 34
  resources for controlling, 35
  standard precautions for, 51–52
  transmission of, 34–35
Informal investigations, 282
Inhalation, 34
Injuries, work-related, 277. *See also* Accidents; Incidents; Workers' compensation
  consequences of, 170–171
  direct costs of, 171
  indirect costs of, 172
  reporting, 279–281
Injury prevention
  holistic approach to, 149–151
  traditional approaches to, 150
Interactive computer-based training, 262
International biohazard symbol, 139(figure)
International Health Care Worker Safety Center, 37
Interviewing, for incident investigations, 284
In the Zone (Mulry), 149
Intimate partner stalkers, 182
Investigations, incident
  formal, 282–283
  informal, 282
  initiating, 281
  interviewing skills for, 284

J

JCAHO. *See* Joint Commission on Accreditation of Healthcare Organizations (JCAHO)
Job hazard analysis, job factors for, 170–171
Joint Commission on Accreditation of Healthcare Organizations (JCAHO), 9, 22, 26–27
  medical management program, 101

K

Keyboards, alternative, 172–173

L

Lab coats, 118
Labor unions, safety and health committees and, 149
Lasers
  beam hazards, 108
  classes of, 108
  nonbeam hazards, 108
  safety guidelines for, 108–109
  safety standards for, 100
Latex glove allergies, 120–122
Lavatories, patient safety in, 213
Learning, creative approaches to, 258. *See also* Safety training
Leggings, 118
Liquid nitrogen, safety guidelines for, 92–93
Lockout, 94–95
Long-term employee training, 265–266

M

Machines, servicing, 94–95
Management
  determining role of, in safety programs, 2–3
  employee involvement and, 147–148
Marine Protection, Research, and Sanctuaries Act (MPRSA), 17
Measurement, for medical equipment safety programs, 106–107
Medical air, safety guidelines for, 91
Medical equipment. *See* Equipment
Medical equipment management, patient safety and, 218
Medical equipment safety programs, 102–103. *See also* Safety programs
  developing inventory stage, 104
  developing management program, 104–106
  implementation stage, 106
  improving, 107
  monitoring and measurement stage, 106–107
  risk assessment stage, 103–104
  training and education stage, 106
Medical gas cylinders, 82–83
  handling and storing guidelines for, 83–84
  leak test procedures for, 84–87
  moving, 87
  opening and closing values, 88
  personal protective equipment for, 84
  receiving procedures for, 84

safety standards for, 83
storage requirements for, 85–86
storing guidelines, 87–88
Medical gases, 82–83
argon, 93
carbon dioxide, 90–91
cylinder safety for, 83
handling and storing of, 83–88
liquid nitrogen, 92–93
medical air, 91
nitrogen and nitrogen NF, 92
nitrous oxide, 88–90
odorizing, 94
oxygen, 91
Medical offices, applying ergonomics in, 162–170
Medical practices. *See also* Health care workers; Safety programs; Workplaces
aging workers in, 267–270
Americans with Disabilities Act and, 21–22
bombs or bomb threats and, 179
building-related illnesses and, 195–200
chemicals in, 61–62
escalating responsibilities of, 1–2
federal regulations for, 13–23
indoor air quality of. *See* Indoor air quality (IAQ)
integrating safety into, 2–3
nongovernmental regulations for, 24–29
patient and staff interaction in, 178
process for establishing safety programs by, 2
protests and, 178–179
relocating, and indoor air quality, 199–200
as safe work environments, 1
state regulations for, 24
threats to personnel of, 179–180
training of multicultural employees in, 273–275
types of equipment in, 101–102
workplace violence and. *See* Workplace violence
young workers in, 269–273
Medical staff personnel. *See* Health care workers
Medical treatments, recording, 294–295
Medical waste. *See* Waste, healthcare
Medical Waste Tracking Act of 1988 (MWTA), 135
MedWatch, the Medical Device Reporting Program (MDR), 18
Mercury, 67–69
Mercury waste, reducing, 134. *See also* Waste, healthcare

Multicultural employees, safety training for, 273–275. *See also* Aging workers; Young workers
Multiple chemical sensitivity (MCS), 71–72
Multiple-Drug-Resistant Mycobacterium Tuberculosis (MDR-TB), 53
Musculoskeletal disorders (MSDs), 160
early identification of, 169
prevalency of, 161
reporting, 168–169
work-related, 278

N

National Center for Chronic Disease Prevention and Health Promotion, 22
National Center for Environmental Health, 22
National Center for Health Statistics, 22
National Center for HIV, STD, and TB Prevention, 23
National Center for Infectious Diseases, 22
National Center for Injury Prevention and Control, 22
National Fire Codes, 27
National Fire Protection Association, 27–28
National Institute for Occupational Safety and Health (NIOSH), 23
National Patient Safety Foundation, 218–219
Natural disasters, 225–226, 233. *See also* Emergency management; Fires
earthquakes, 235–236
floods, 233–234
hurricanes, 238–239
severe winter storms, 234–235
tornadoes, 236–238
Needles, disposing of, 40
Needlestick injuries. *See also* Infections; Sharps injuries
preventing, 37, 39–40
prevention programs for, 38
Needlestick Safety and Prevention Act, 40–41
Needs assessments, for safety programs
applying, 8
elements of, 5–8
preparing for, 4–5
New drug applications (NDAs), 79
New employee orientation level of training, 264–265
NFPA 101 Life Safety Code, 27
Nightingale, Florence, 1
Nitrogen, safety guidelines for, 92
Nitrogen NF, safety guidelines for, 92
Nitrous oxide, safety guidelines for, 88–90

Nongovernmental organizations. *See also* Federal agencies; Government compliance agencies
American Conference of Governmental Industrial Hygienists, Inc., 25
American Nation Standards Institute, 28
American Society of Heating, Refrigerating, and Air-Conditioning Engineers, 25–26
Compressed Gas Association, 28–29
Joint Commission on Accreditation of Healthcare Organizations, 26–27
National Fire Protection Association, 27–28
Nuclear Regulatory Commission, 20–21

O

Objectives, safety program, 3
Occupational exposure values61, 59–61
Occupational health and safety, 13
Occupational injuries. *See* Injuries, work-related
Occupational medicine, 13
Occupational Safety and Health Administration (OSHA), 13, 22
bloodborne pathogens standard, 41–49
emergency planning regulations, 240–245
ergonomics and, 174
General Duty Clause, 15
Hazard Communication Standard, 15, 73–77
hazardous chemicals and, 16
inspections by, 15
recordkeeping guidelines, 294–300
State Plan states and, 15–16
workplace violence and, 190–191
Odorization of atmospheric gases, 94
Office of Justice Programs (OJP), 19–20
OSHA. *See* Occupational Safety and Health Administration (OSHA)
OSHA 101 form, 296–299
OSHA 200 log, 204–296
Outcomes, safety program, 3
Overt workplace injuries, 161
Oxygen, safety guidelines for, 91

P

Parking areas, patient safety in, 209
Parts of substance per million parts of air (PPM), 61
Patient care equipment. *See* Equipment
Patient safety
in building entrances, 209–210
checklist for developing programs for, 207–208
children and, 215–217
emergency management and, 218

INDEX 313

in examination rooms, 212
in exits, 213
falls and, 214–215
in hallways and aisleways, 211
housekeeping practices for, 213
in lavatories, 213
medical equipment management and, 218
in office entrances, 210
outside of medical facilities, 208–210
in parking areas, 209
persons with mental disabilities and, 217
persons with physical disabilities and, 217–218
in preexamination rooms, 211–212
senior citizens and, 214–215
in storage areas, 212–213
in waiting areas, 210
Permissible Exposure Limit (PEL), 60–61
Personal health, integrating, with healthful workplaces, 149–153
Personal protective equipment (PPE), 156
　for bloodborne pathogens, 47–48, 49–50
　for chemotherapeutic drugs, 63–64
　cleaning and maintenance of, 119
　comfort and fit for, 114
　for emergencies, 244
　employee training for, 119–120
　for formaldehyde, 66
　for glutaraldehyde, 67
　hazard assessment and selection of, 111–114
　for medical gas cylinders, 84
　for mercury, 68
　overcoming employee complaints for, 114
　for phenolics, 69–70
　protective devices for, 114–119
　recordkeeping for, 120
Persons with mental disabilities, patient safety and, 217
Persons with physical disabilities, patient safety and, 217–218
Pest control, for improving indoor air quality, 201
Pharmaceutical drug abuse, 80
　sample warnings for, 80–81
　solutions, 81–82
Pharmaceutical drugs, 79–80
Phenolics, 69–70
Pollution Prevention Act (PPA), 17
Preexamination rooms, patient safety in, 211–212
Priorities, setting, 9
Procurement planning, for equipment, 102
Promotional programs, 153

Protests, 178–179
Public Interest Research Groups (PIRG), 215
Publicly owned treatment works (POTW), 131

R

Radioactive waste, 20
Ramazzini, Bernardino, 13
Ramps, patient safety in, 209
Recognition programs, 154
Recommended Exposure Limit (REL), 61
Records
　employee attendance, 10
　for health and safety programs, 9–10
Regulation, Principles of Good, 20
Regulatory inspections, guidelines for, 29–31
Repetitive stress injuries (RSIs), 160
Research and Special Programs Administration (RSPA), 19
Resource Conservation and Recovery Act (RCRA), 16
Respiratory protection, 122–123
Reward programs, 152
Right-to-Know Law. *See* Hazard Communication Standard (HAZCOM)
Risk management, for accidents, 290–291
Ritalin, 79–80
Role-playing, as training technique, 260

S

Sabotage, 176. *See also* Workplace violence
Safe Drinking Water Act (SDWA), 16
Safe Medical Device Reporting Act (SMDR) of 1990, 293
Safe Medical Devices Act (SMDA) of 1990, 17–18, 99–100, 299–300
Safety, communicating commitment to, 10–11
Safety and health committees
　establishing, 148–149
　functions of, 149
　labor unions and, 149
　recordkeeping and, 10
Safety and health programs
　control methods for, 154–157
　for families, 154
　promotion and recognition programs for, 153–154
Safety coordinators, responsibilities of, 9
Safety hazards. *See* Hazards
Safety programs. *See also* Medical equipment safety programs
　addressing diverse needs and, 3–4
　annual evaluations for, 303–304
　benchmarking, 305

　commitment and, 3
　conducting needs assessments for, 4–8
　core elements for successful, 2–3
　determining role of management and, 2–3
　developing policies and procedures, 9–10
　establishment of, by medical practices, 1
　evaluating, for resource allocation, 304–305
　evaluating existing, 6–7
　evaluating for new and revised, 304
　evaluation process for, 301–303
　role of management in determining, 6–7
　setting priorities for, 9
　tracking accidents and illnesses, 277–278
Safety records, 10
Safety training
　auditing retention of, 275
　current employee level of, 265
　distance learning for, 262
　fundamentals of, 256–257
　generation-specific, 267–273
　for long-term employees, 265–266
　for multicultural employees, 273–275
　for new employees, 264–265
　regulatory agency requirements for, 257
　self-directed learning, 262–263
　supervisors and, 266–267
　techniques of, 257–261
Safety training programs, 248–249. *See also* Employee training
　anticipating additional needs for, 254
　conducting instructor-led, 250
　confirming planning responsibilities, 250
　designing, 250
　establishing needs, 249
　evaluating, 250–254
　evaluation forms for, 251–252
　selecting type of, 249–250
　setting up, 247–248
　specialty practices considerations for, 254–256
Segregation, of medical waste, 137
Self-directed learning
　advantages of, 262
　change agents in, 261–262
　disadvantages of, 262–263
Self-study, as training technique, 260–261
Semmelweis, Ignaz Philipp, 1
Severe winter storms, 234–2335
Sharps, disposing of, 40
Sharps injuries, 38. *See also* Needlestick injuries
Short-Term Exposure Limit (STEL), 61

Sidewalks, patient safety in, 209
Slips, preventing, 96
Smoking, and indoor air quality, 202
Snow storms, 234–235
Sprains, 149
Stairs, patient safety in, 209
Stalkers, types of, 182
Stalking, 180–182
State Plan states, 15–16
State regulations, 24
Storage areas, patient safety in, 212–213
Strains, 149
Strategic classrooms, 263
Stress
  effects of, 151–152
  solutions to, 153
  sources of, 151
  in workplaces, 152
Substitution methods, 155–156
Suicides, 176. *See also* Workplace violence

T

Tagout, 94–95
Tasks, and workstation interrelationships, 170
Technological emergencies, 239. *See also* Emergency management
Threats, to health care workers, 179–180
  checklist for, 181
Threshold Limit Value (TLV), 61
Time-weighted average (TWA), 61
Tornadoes, 236–238
Torso and additional body protection, 118
Toxic Substances Control Act (TSCA), 17
Toy safety, in waiting rooms, 215–216
Training. *See* Employee training; Safety training
Training techniques, 257–261
Trips, preventing, 96
Tuberculosis (TB). *See also* Infections
  infection control measures for, 54–56
  Multiple-Drug-Resistant Mycobacterium, 53
  occupational exposure to, 52–53
  respirator requirements, 56
  respiratory protection for, 123

U

Unions, safety and health committees and, 149
Universal precautions, for bloodborne pathogens, 44
Uranium Mill Tailings Radiation Control Act (UMTRCA), 17
US agencies. *See* Government compliance agencies

V

Values, human, 2–3
Valves, cylinder, opening and closing, 88
Vengeful stalkers, 182
Videoconferencing, 26
Violence. *See* Domestic violence; Workplace violence
Violence Against Women Act (VAWA), 19–20. *See also* Workplace violence
Virtual universities, corporate, 263
Vision, establishing, 2

W

Waiting areas
  childproofing, 217
  patient safety in, 210
  toy safety in, 215–216
Waste, healthcare
  CDC definition of, 135–136
  choosing disposal contractors for, 143–146
  conducting survey for, 128
  containment of, 137–138
  contingency planning for, 141–153
  Council of State Governments guidelines for, 136–143
  determining general status of, 128–129
  disposing of, 130–131
  emergency plans for, 132
  EPA categories of, 135
  handling, 137
  identifying, 127–128
  labeling, 130, 139
  liability issues, 126–127
  monitoring and recordkeeping of, 141
  planning programs for, 136
  process of, 126
  reducing, 132–133
  segregating, 129, 137
  storage prior to treatment of, 140–141
  storing, 129–130
  training for handling, 141
  types of, 125–126
Web-based training, 262
Winter storms, 234–235
Workers' Compensation, 285
  costs, 287–289
  principles of, 287
  returning employees to work and, 289–290
Workplaces. *See also* Medical practices
  control methods for reducing hazards in, 154–157
  ergonomics in, 161–162
  integrating healthful, with personal health, 149–153
  stress and, 152
Workplace safety. *See* Occupational Safety and Health Administration (OSHA)
Workplace violence. *See also* Domestic violence
  definition of, 176–177
  environmental risk factors for, 184–185
  human risk factors for, 186–188
  organizational after incidents of, 188–189
  organizational risk factors for, 186
  OSHA and, 190–191
  prevention program for, 189–190
  risk factors for health care and social workers of, 175–176
  types of, 177–183
Work practices
  for chemotherapeutic drugs, 63
  controls for bloodborne pathogens, 47
  for glutaraldehyde, 66
  for mercury, 68
  for reducing workplace hazards, 156
Work-related musculoskeletal disorders (WMSDs), 278. *See also* Musculoskeletal disorders (MSDs)
Workstations, and tasks interrelationships, 170

Y

Young workers. *See also* Aging workers; Multicultural employees
  motivation factors for, 271–272
  needs of, 269–271
  rights and responsibilities of, 271
  safety concerns regarding, 272–273
Zoonotic transmission, 35